Digital
Satellite Service

Other Books by Robert L. Goodman

Maintaining and Repairing VCRs—Fourth edition
Practical Troubleshooting with the Advanced Video Analyzer
The TAB Service Manual for CCTV and MATV
Troubleshooting and Repairing Digital Video Systems: TVs and VCRs,
Microprocessors, Signature Analysis
Troubleshooting and Repairing Electronic Circuits—Second edition
Troubleshooting with Your Triggered-Sweep Oscilloscope

Digital
Satellite Service
Installation and Maintenance

Robert L. Goodman

TAB Books
Imprint of McGraw-Hill

New York San Francisco Washington, D.C. Auckland Bogotá
Caracas Lisbon London Madrid Mexico City Milan
Montreal New Delhi San Juan Singapore
Sydney Tokyo Toronto

McGraw-Hill

A Division of The McGraw·Hill Companies

©1996 by The McGraw-Hill Companies, Inc.

pbk   1 2 3 4 5 6 7 8 9   FGR/FGR   9 0 0 9 8 7 6

Library of Congress Cataloging-in-Publication Data
Goodman, Robert L.
   Digital satellite service : installation and maintenance / by
Robert L. Goodman.
      p.   cm.
   Includes index.
   ISBN 0-07-024205-4 (pbk.)   ISBN 0-07-024204-6 (hard)
   1. Direct broadcast satellite television.   I. Title.
TK6677.G66   1996
621.388'53—dc20   96-648
                                                                   CIP

McGraw-Hill books are available at special quantity discounts to use as premiums and sales promotions, or for use in corporate training programs. For more information, please write to the Director of Special Sales, McGraw-Hill, 11 West 19th Street, New York, NY 10011. Or contact your local bookstore.

Acquisitions editor: Roland S. Phelps
Editorial team: Robert E. Ostrander, Executive Editor
                Joann Woy, Indexer
Production team: Katherine G. Brown, Director
                Ollie Harmon, Coding
                Brenda M. Plasterer, Coding
                Lisa M. Mellott, Desktop Operator
Design team: Jaclyn J. Boone, Designer
                Katherine Lukaszewicz, Associate Designer

0242054
EL3

Acknowledgments

A book that covers the magnitude of the very technical, complicated Digital Broadcast System could not have been accomplished without the assistance, photos, and information of many companies and their personnel.

So, a GREAT BIG THANK-YOU to the following companies and their helpful folks:

Thomson Consumer Electronics (RCA DSS):
Mr. Scott Stevens and Mr. Tom Graff

GM Hughes Space and Communications Company:
Ms. Fran Slimmer, Chief, Media Relations

DirecTV Uplink, Castle Rock, CO:
Ms. Lynn Hill, Public Relations Specialist

United States Satellite Broadcasting, Inc. (USSB)
Uplink, Oakdale, Minnesota (Hubbard Broadcasting):
Stanley S. Hubbard, Chairman and CEO
Stanley E. Hubbard, President
Robert W. Hubbard, Executive Vice President
Mary Pat Ryan, Senior Vice President, Marketing
Ray Conover, Vice President, Engineering
John Degan, Vice President, Operations
Ralph Dolan, Vice President, USSB Programming

Thanks Again
Bob Goodman, CET
Hot Springs Village, Arkansas

This book is dedicated to
"Gloria,"
a genuine southern lady of the first class.

Contents

Introduction

THIS BOOK CONTAINS THE INFORMATION YOU SHOULD KNOW before signing up for these new, small-dish DSS satellite TV reception systems. It will help you to wisely select a system for your needs and show you how to install and adjust your own receiver and dish.

Information, photos, and drawings will be included for your assistance. Electronics technicians will also benefit from this book, by using it to determine if they should undertake training for the installation of the hardware for this new and fast-growing entertainment medium.

One chapter will be devoted to the "SMATV" distribution system. The acronym SMATV stands for "satellite master antenna television" system. This system combines the DSS satellite dish signal with off-air (terrestrial) or standard TV broadcast signals for a complete "in-house" video distribution network in the home.

Information on the Hughes satellite, sometimes referred to as the "high-powered bird," will be included in this book. The first DBS satellite was launched in December of 1993, and a second one went up in mid-1994.

Each satellite has sixteen high-powered (120-watt) transponders. They have been designed to accommodate analog and digital signals. They also have the capability to receive 16×9 wide-screen TV signals, as well as HDTV broadcasts once that revolutionary technology is in place.

The digital signals will be decoded by an integrated receiver/descrambler, which will reproduce them with "near laserdisc" quality. The stereo sound you will hear is dramatic CD-quality audio.

Also included will be a list of programs available for your viewing pleasure at the time of this book's publication. As more programs go online with the DSS system, all you "channel surfers" and "remote clickers" should have lots of fun viewing time.

Read, view, and enjoy.

Satellite TV history

IN THIS BEGINNING CHAPTER WE WILL LOOK AT A BRIEF history of the "big dish" format originally used for analog TV satellite transmissions; then we'll take a quick look at the new "big bird" in the sky, digital satellite systems and the world-wide satellite delivery system.

Basic satellite operation

TV communications satellites generally orbit the earth at the same relative rate of speed as the earth's rotation. Satellites in such an orbit appear to remain fixed in relation to a specific point on the earth, such as your home receiving dish, thus enabling satellites to relay uninterrupted signals from one point on the surface to another.

A TV satellite signal originates in a television studio control center. The signal is fed to an "uplink" dish that beams the signal up to the orbiting satellite. The satellite then beams the signal back down to earth, where a "downlink" system (an antenna dish) receives it, then routes the signal through amplifiers, a receiver, and a decoder. The strength of the signal in this "footprint" is not the same for all locations. Most of the older "big dish" analog systems are aimed at the midwestern U.S.

Simplistically speaking, a satellite television system consists of an antenna (dish), a low noise block down-converter (LNB), a receiver, and a TV set. The antenna dish is made of reflective material that gathers and focuses the signal to the feedhorn at the center of the dish. These first analog satellite receiving systems are in the 3.7-GHz frequency band, and have to listen to very low-power birds. Thus a large dish is required, and it has to be repositioned for each satellite that you wanted to receive programs from.

Because the signals are quite weak and must be amplified, the LNB (which is located in the feedhorn) must magnify the signal

100,000 times without generating undue quantities of thermal "noise," which will appear on the TV screen as snow.

The signal is down-converted and forwarded onto the receiver, which "decodes" the signal, transforming it into the original broadcast TV picture and sound. This in turn is fed into a channel of your TV set.

Twice a year, in October and March, the sun passes directly behind each of the satellites. This solar blackout is noticeable for approximately two weeks.

At first the interference will be very slight; however, during the peak day (halfway through the two-week cycle) the interference from the sun will completely black out all of the signals from the satellite directly in front of it for approximately 10 minutes.

The following is a list of some "C" band satellites in geosynchronous orbit:

Satellite	Designation
Satcom 1	F1R
Galaxy 1	G1
Satcom 3	F3
Telstar 303	T3
Westar 5	W5
Spacenet 1	S1
Morelos 2	M2
Morelos 1	M1
Anik 1	D1
Westar 4	W4
Telstar 301	T1
Galaxy 3	G3
Westar 3	W3
Telstar 302	T2
Satcom 4	F4
Westar 2	W2
Comstar 3/4	D3/D4
Galaxy 2	G2
Satcom 2	F2

One of the early makers of home satellite dish receiver equipment was Houston Systems, Inc., located in Houston, Texas. Their system was called the Tracker System V Receiver/Positioner. A drawing of the complete setup for the Tracker System is shown in Fig. 1-1 with the proper grounding points.

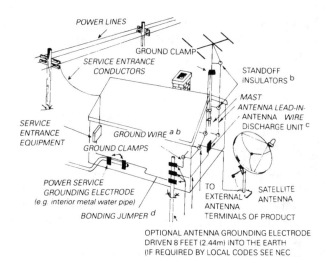

POWER LINES

GROUND CLAMP

SERVICE ENTRANCE CONDUCTORS

STANDOFF INSULATORS [b]

MAST

ANTENNA LEAD-IN-ANTENNA WIRE DISCHARGE UNIT [c]

SERVICE ENTRANCE EQUIPMENT

GROUND WIRE [a] [b]

GROUND CLAMPS

POWER SERVICE GROUNDING ELECTRODE (e.g. interior metal water pipe)

BONDING JUMPER [d]

TO EXTERNAL ANTENNA TERMINALS OF PRODUCT

SATELLITE ANTENNA

OPTIONAL ANTENNA GROUNDING ELECTRODE DRIVEN 8 FEET (2.44m) INTO THE EARTH (IF REQUIRED BY LOCAL CODES SEE NEC SECTION 810-21 (1)

■ **1-1** *A set-up drawing of the Tracker Satellite System with all proper grounding points.* DirecTV, Inc. Thomson Consumer Electronics

The home TVRO system

The home satellite TV system consists of five major components. The first is the TV studio and master control facility, from which the programmer originates the satellite TV feeds. This programming is then delivered to the transmit-receive (T-R) earth station via videotape or by direct video circuit. The signal is uplinked to the geosynchronous communications satellite in the 6-GHz band. The satellite, often called the "space segment," electronically amplifies this incoming signal, shifting it downward in frequency to the 3.7-GHz band, and retransmits it earthward. On the viewer home front, TVRO earth station terminals scattered throughout the countryside simultaneously receive the satellite feed. A master satellite command earth station, operated by the satellite common carrier owner, monitors the "housekeeping" functions of the satellite such as battery charge, orbital positioning, etc., and controls the overall operation of the space satellite.

The studio and master control facility of the programmer is often part of an affiliated terrestrial TV station. As an example, Trinity Broadcasting Network uplinks live program feeds from their Los Angeles UHF station to form the basis of the TBN network. Programs can originate from separate studio and control facilities, as is the case with the Cable News Network. They can also be uplinked by common carriers at the earth stations, using banks of

3

videotape players to play back prerecorded tapes. The HBO movie service has used the RCA American New Jersey uplink facility for this purpose.

The T-R uplink earth station consists, typically, of a 30-foot parabolic dish antenna and associated transmitter electronics. The uplink is often owned by the common carrier, which also owns the satellite system. The common carrier T-R station receives the incoming video signal, which is then processed for its trip to the satellite.

At the home receiver, a low-noise amplifier is mounted on the end of the feedhorn. This is an extremely sensitive high-frequency transistor circuit that can amplify the very weak signals received from the low-power satellite without swamping them with excessive noise generated in the amplifier itself. Because the microwave signal coming from the satellite transponder is so very weak (these satellite transponders have a power output of 5 to 10 watts), it calls for a lot of preamplification. The received signal is less than one-millionth as strong as the local common-carrier repeater output. At these low signal levels, even the natural noise produced by the radiated heat from the earth into space can interfere with the far-out satellite signals. LNAs are therefore rated in terms of their noise figure, represented by Kelvins (K).

Noise, then, is the predominant characteristic of interest of the amplified signal. The actual amount of noise the LNA generates is compared with that of a device operated at zero Kelvin (minus 270 degrees Celsius). The output of the LNA is fed through a special low-loss, large-diameter coaxial cable (call "Heliax") into the "down-converter." The down-converter shifts the 3.7-GHz signal to an intermediate frequency (IF) signal usually in the 70-MHz range. This IF signal can then be carried over inexpensive standard coaxial cable.

In the early days, down-converters were located at or inside the satellite receiver, requiring the bulky and costly Heliax cable to bring the 3.7-GHz TV signal from the LNA to the receiver. Most cable TV systems still use this type of installation. However, it is now possible to mount the down-converter at the antenna site. Now, many TVRO satellite system manufacturers use the "dual conversion" technique by having a separate waterproof down-converter electronics package mounted near the LNA. Some companies have now combined LNA/down-converters in a very small package. This device, called a low-noise converter (LNC), allows low-cost coax cable to be run for hundreds of feet from the antenna to the home satellite receiver.

4

The satellite receiver is the most important component of the home satellite TV terminal. Unlike a TV set, the satellite receiver does not have a built-in audio amplifier or picture tube, and is more like a stereo component unit than a TV. The satellite receiver takes the 3.7-GHz or IF TV signal (depending on whether the down-converter is located at the antenna or built into the receiver itself) and converts the signal to baseband video and audio, which is outputted at rear of the receiver. The baseband video signal cannot be fed directly into a home TV set. The signal is similar to that produced by a video camera output jacks. To view this video on a TV receiver, it must first be fed into an RF modulator to convert the signal to a VHF TV channel. The output of the RF modulator is then fed through a lead-in cable to the VHF antenna terminals on the back of a TV set, and the TV is then tuned to channel 3 or 4. This works in the same way that a VCR signal is fed into a TV set for viewing tapes.

The baseband video output of the satellite TVRO receiver can be fed directly into a commercial TV monitor or projection TV system, or into a VCR unit (which eliminates the need for a modulator). This setup improves the fidelity of the picture because the modulator, with its added noise and distortion, can be eliminated from the loop. Most TVRO receivers provide built-in VHF modulators, and the audio output of the satellite receiver is heard through the TV speaker.

Many companies have manufactured satellite TV receivers over the years. One of the first for the home TVRO market was the International Crystal unit, first produced in 1979. This receiver tuned all 24 channels on the transponder birds, and allowed the user to select between the various audio subcarriers which appeared along with the video picture. Many satellite programmers transmit up to 10 or more additional audio channels along with the TV program audio. By varying the audio subcarrier tuning control on the satellite TVRO receiver, these other audio channels may be heard as well.

Twenty-four transponder communications satellite utilize a frequency reuse technique known as "polarization." Twelve channels are transmitted using microwave signals that are polarized vertically with respect to the positioning of the satellite. A second group of 12 channels is transmitted using microwave signals that are polarized horizontally. These wave groups formed are orthogonal, or at right angles to each other, and they do not interfere with each other; this allows the 24 channels to be overlapped in frequency. Thus, each satellite will consume only half as much bandwidth as would otherwise be needed to transmit 24 simultaneous channels.

Back on earth, the TVRO feedhorn/LNA combination is rotated along its axis 90 degrees to pick up the separate vertical and horizontal 12-channel transponder groups. This is often accomplished by using a small TV antenna rotor. There is also a "polar rotor" electronic polarizing device that performs this same function by rotating just the front end of the feedhorn probe.

Most commercial TVRO satellite terminals solve this problem another way. They use two separate LNAs and cable feeds, which are mounted on a special feedhorn that splits the signals into their corresponding vertical and horizontal components. The 12 horizontal transponders are shifted upward in frequency by 20 MHz (one-half transponder) from their next-door vertical neighbors, but they all share the same 3.7-GHz band. Thus, any satellite TVRO receiver is capable of tuning the full 24 satellite transponder set by providing a fine-tuning "frequency offset" control on the front panel of the receiver.

6   Commercial satellite systems

There are now all kinds of satellites twirling around in space. There are satellites that look at weather patterns, relay telephone and computer data, and feed us all kinds of video information for education and entertainment.

Restaurants, motels, and hotels receive movies and other music and entertainment via the satellite dish. And of course hospital, church, and school organizations can all use satellite television for information, education, and instruction purposes. In addition, the hospital, like a hotel, can provide first-run movie service to its patients via satellite TVRO earth station, helping to entertain patients during their confinement. Many hotels generate a profit from the equipment by providing an in-room pay-movie service that is charged on a pay-per-view basis.

The movie channel provider usually installs a CATV-like digital decoder adjacent to each television set; this unit works in conjunction with the central billing computer. When any room guest selects a desired movie, the billing will be automatically posted to the room account via the special equipment. The movies themselves are provided on videotape cassettes that are continuously running in a multidrive video playback system located in the hotel.

Other hotels may take a single movie channel feed delivered via an independent MDS common carrier, which provides the programming to the hotel complex via terrestrial microwave service. These hotels have microwave dish antennas that are mounted on the roof and aimed at the local MDS transmitter, which is located on a nearby hill or building top. The pay movie channel is converted by a small down-converter electronics package into a VHF television channel, which is then fed into the hotel's master antenna system, as shown in Fig. 1-2. Some hotels that provide this type of movie service usually do so on an inclusive basis, offering the movies free as part of the room charge.

DirecTV, Inc. Thomson Consumer Electronics

■ **1-2** *The RCA DSS 18-inch satellite dish mounted on the side of a building.*

The hotel management that installs its own TVRO earth station can obtain movies directly from the satellite, thus bypassing the local suppliers. Other satellite-fed channels can be distributed as well, increasing the overall quality and value of the hotel room television service. For example, the Cable News Network (CNN), the Financial News Network, and C-SPAN can provide the professional and business guest with information services. The children's networks can be carried to entertain the youngsters, while the sports channels can capture the avid sporting fan.

The TVRO earth station can also be used to plug the hotel into the rapidly expanding videoconferencing networks. Many companies now produce and distribute via satellite national video teleconferences, political fund-raisers, and Fortune-500 annual meetings. Special events and sporting activities like championship boxing matches can be carried on a pay-per-view basis to the guest rooms and public-admittance areas of the hotel. The Reagan Administration often used videoconferencing to bring together thousands of people at gala fund-raising parties held in hotels throughout the country. The president could reach out to each of these locations electronically via satellite; two-way audio talkback channels were usually provided to allow people at the various hotels to be heard back in Washington, D.C.

Charity events can be staged in hotels with satellite videoconferencing facilities. Closed-circuit meetings can increase the room registrations, while saving the participants time and travel expenses. For the major corporations, annual video shareholders' meetings and video teleconferences tying together the regional sales offices can save hundreds of thousands of dollars.

Many of the national hotel chains are now installing their own TVRO earth stations. The first and largest of these, the Holiday Inn group, now has several hundred five-meter dishes strategically located at hotels throughout the country. The Hilton, Hyatt, and Marriott chains are following this lead, and most national hotel chains will soon have their own TVRO earth stations installed and operating. Some now have plans to develop their own private channels to present top entertainment acts originating in their Las Vegas and Los Angeles locations for nationwide satellite distribution.

Satellite video teleconferencing

Video teleconferencing is one of the most significant business tools to have been spun off the NASA space program technology. The development of the communications satellite has slashed the cost of providing a nationwide multipoint video channel. Before the advent of the domestic communications satellites, only the television networks could afford to lease the terrestrial television circuits provided by AT&T on an occasional basis. The costs and infrastructure required to fully utilize this technology were just too great. Today, hundreds of hotels, hospitals, office buildings, and theater complexes are equipped with TVRO satellite terminals. Many major organizations have appeared to provide "turnkey" videoconferencing services. Users have included GM Hughes,

TRW, Ford Motor Company, The American Bar Association, and many more.

Videoconferencing is a big industry and will grow much more. Some use a split-screen technique and have a second screen that can also display graphics. Looking at the increased revenues that videoconferences can generate, most of the major hotel chains have committed themselves to installing their own TVRO and associated video teleconferencing facilities throughout the country. The Holiday Inn Hi-Net System has stations in many cities and almost all the states in the USA. The Ramada Inn chain has equipped a number of its hotels with TVROs, and the Hilton, Sheraton, and Hyatt Hotel conglomerates are now installing their own networks. Also, some conventions have set up temporary satellite installations.

Conferencing services

Videoconferencing services are now used for just about any purpose by corporations, nonprofit organizations, and government agencies in every state. There are many major functions for a videoconference, and some of these are listed below:

1. Convention and trade show meetings
2. Education conferences and symposiums
3. News conferences
4. General business meetings
5. Stockholder meetings
6. Sales and marketing meetings (both national and international)
7. Fund-raising events for political parties
8. Charitable events and telethons

Satellite communications

The earth is surrounded by artificial satellites today. Many of these satellites carry repeaters and are used for communications. In recent years satellites have been placed in synchronous orbits, providing continuous intercommunications capability among almost all locations on the world's surface.

In the 1960s, a series of passive satellites was launched into orbit around the earth. These were large metalized balloons that reflected radio waves sent up to them. The Echo satellites were placed in low orbits, because the existing booster rockets could not lift a satellite into a synchronous orbit. The area of coverage for each Echo satellite was limited by the low orbit, and access time was very brief.

The active communications satellite was developed after the Echo satellites. An active communications satellite is an orbiting repeater with broadband characteristics. The signal from the ground station is intercepted, converted to another frequency, and retransmitted at a moderate power level. This provides much better signal strength at the receiving end of the circuit as compared with a signal reflected from a passive satellite. The first active satellites were placed in low orbits and had the same shortcomings as the previous Echo satellites. Finally, active communications satellites were placed in synchronous orbits, making it possible to use them with fixed antennas, with moderate transmitter power, and at any time of the day or night.

Synchronous satellites are used for television and radio broadcasting, communications, weather forecasting, and military operations. Telephone calls are now routinely carried by satellite.

Satellite system

In electronics the term satellite has two meanings:

1. An unmanned spacecraft in orbit.
2. An electronics system, such as a transmitter or power station, that is subordinate to or dependent on a larger or more complete electronic system.

Spacecraft launchings

Satellites are lofted into orbit by rocket propulsion to perform one or more functions. Most are specialized for:

1. Relay of terrestrial communications.
2. Active or passive electromagnetic surveillance, photographic reconnaissance, or television reconnaissance.
3. Navigation aids.
4. Unmanned exploration of space by instrumentation.
5. Monitoring of physical phenomena.
6. Scientific experimentation.

A satellite system includes both the spacecraft and ground-based support systems for control and communications, usually by radio link. However, the signals from many broadcast and navigation satellites can be received by large numbers of suitable directional or omnidirectional antennas distributed over a wide geographical region.

Some satellites are put in low orbit to travel around the earth on a predetermined trajectory at scheduled times. Other satellites can be placed in high or geostationary orbits so they remain over fixed points on the earth's surface. Space probes, however, are usually lofted out of a low orbit into an interplanetary trajectory.

Military reconnaissance or surveillance satellites are usually placed in low orbit so they pass repeatedly over regions of the earth of interest to the nations that have launched them. They might be equipped with radar, infrared sensors, broadband radio receivers, photographic cameras, television cameras, and various combinations of these.

Commercial satellites in low orbit can use infrared sensors and photographic or television cameras. They can monitor and map changes on the earth, in the sea, and in the atmosphere. Scientific satellites monitor weather patterns, magnetic fields, and changes in radiation strength. Navigation satellites transmit coded signals that permit aircraft and ships to establish their positions with respect to latitude and longitude.

11

Most active satellites are powered by solar cells that convert the sun's radiation directly into electric power. The solar cells can be mounted on the outside of the cylindrical satellite body, which is continuously rotated so the solar cells always face the sun while the antenna is kept pointed at the earth. The spinning body acts as a gyroscope to stabilize the satellite in space.

Other satellites are stabilized with built-in gyroscopes in both the body and large solar panel arrays. This satellite configuration can provide more solar power than the spin-stabilized units, because larger solar panels with larger areas can be positioned so they are continuously illuminated by the sun.

Most satellites have one or more antennas that receive signals from the earth and transmit them back at another frequency. Commercial communications satellites have earth stations that transmit signals to the satellites on uplink frequencies of 6 or 14 GHz. The received information is amplified and relayed back to earth on the lower downlink frequencies of 4 to 12 GHz. U.S. military communications satellites use uplink frequencies of 8 GHz and downlink frequencies of 7 GHz.

Most satellites, regardless of function, are fitted with instruments that measure such variables as temperatures, radiation, and magnetic field in and around the satellite for transmission back to their earth stations. Some also carry receivers for signals that control onboard jets to make orbital corrections due to satellite drift.

Because most satellites travel in programmed orbits, only the small differences between the data derived from the actual orbit and the program have to be monitored and corrected. By contrast, other satellites and space probes are equipped with star-tracking systems for automatic orbital guidance. Sensors within the systems track the sun or designated navigational stars.

The major milestones in satellite technology include the Soviet Sputnik, orbited in 1957, and the two American weather satellites, Tiros and Nimbus, launched in 1964. Important early communications satellites include Telstar (1962), Relay, Syncom, Echo, and Early Bird. They established the technology that made it possible to transmit live television broadcasts across the Atlantic. The first communications satellite, Echo, was a passive reflector of radio signals, but Telstar and all communications satellites that followed have been active repeaters. Scientific exploration of interplanetary space started with the American Ranger and Mariner and Russian Lunik space probes. The ground tracking stations for space probes and communications satellites have highly sensitive antennas with typical gains of 60 decibels (1 million times) for receiving weak signals. The tracking antenna may be rotated with a directional accuracy of about one-thousandth of a degree (3.6 seconds of arc).

The TDRS (tracking and data relay satellite) communications satellite has a hexagonal body that is nine feet across, and it weighs 4,668 pounds. The satellite is equipped with seven antennas, the largest of which are parabolic antennas with a 16-foot diameter. Two solar panels produce 1,700 watts of power.

A module in the lower half of the hexagonal body houses the communications equipment, and the payload module in the upper half contains the electronic equipment for communications with other spacecraft. An attitude control subsystem stabilizes the TDRS and keeps its antenna and solar panels properly oriented.

These communications satellites are in circular geosynchronous orbits at a distance of about 22,250 miles (35,800 kilometers) from the center of the earth.

Also, the planes of their orbits must contain the equator. Because the periodic times of these orbits are 24 hours, they are synchronous with the earth's rotation. At the geosynchronous altitude a satellite can see about 120 degrees of latitude, or about 40 percent of the earth's surface.

Satellites circling the earth in orbits that are lower than the geosynchronous orbit move faster than the earth. As a result, they will appear to rise and set from any point on the earth. All transmitting and many receiving antennas used with these satellites must be able to track them. Any system based on low-orbiting satellites must have multiple satellites for global coverage. An example is the global positioning system (GPS), which will include 21 Navstar navigation satellites for continuous global coverage when it is completed.

Satellite radio repeaters

A repeater is a radio system that picks up and then retransmits a signal to provide long-distance communications. Repeaters are usually used at VHF, UHF, and microwave frequencies. Repeaters are especially useful for mobile operation. The effective range of a mobile station is greatly enhanced by a repeater.

A radio repeater consists of an antenna, a receiver, a transmitter, and an isolator. The transmitter and receiver are operated at slightly different frequencies. The separation is approximately 0.3 percent of the transmitter frequency. This separation of the receiver and transmitter frequencies allows the isolator to work at maximum efficiency, preventing undesirable feedback.

Repeaters are placed aboard satellites. All active communications satellites use repeaters. A satellite in a synchronous orbit can provide coverage over approximately 30 percent of the globe.

High-power direct TV broadcast

A new DBS (Direct Broadcast Satellite) is now in geosynchronous orbit and is beaming digital TV programs to 18-inch-diameter dishes. The digital signals are decoded by an integrated receiver/descrambler, which reproduces pictures of near-"Laserdisc" quality. The programming (which will be available from two or more suppliers) will consist of up to 150 channels of cable network programs, movies, sports, and pay-per-view events. Audio music with "compact-disc quality" will also be available. The Digital Satellite System or DSS is now in operation in most sections of the United States. Two satellites of this DSS system are now in orbit and operational.

A number of companies have gone together and invested over one billion dollars in this satellite system venture. The main investor is Thomson Consumer Electronics, which is building the receivers under its RCA brand name. The photo in Fig. 1-2 shows the 18-inch RCA DSS receiving dish mounted on the side of a building. Thomson Consumer Electronics has developed the digital compression technology for this DSS system. GM Hughes Space and Communications company has built and launched two of the DSS satellites. DirecTV, Inc., a unit of GM Hughes Electronics, is one of the TV program distributors. United States Satellite Broadcasting or USSB, a subsidiary of Hubbard Broadcasting, Inc. is the other program distributor at this time.

A new DBS kid on the block

A new Direct Broadcast Satellite system was launched in the fall of 1995, adding the first direct competitor to the Hughes/Hubbard DirecTV service that is now operating.

The FCC recently approved the sale of SSE Telcom's DirectSat Corp. to ECHOSTAR Communications. The acquisition means that the construction permit and orbital slot assignment for 119 degrees west, a prime slot for U.S. coverage, are now the control of a single entity. ECHOSTAR has stated it intends to move ahead with its DBS plans, using 22 frequencies from that location.

"ECHOSTAR's goal is to offer a competitive DBS service," said Carl Vogel, chief operating officer of ECHOSTAR. "ECHOSTAR is well on the way to making our goal a reality."

The company's first satellite, EchoStar I, was constructed by Martin Marietta and was launched in the fall of 1995. EchoStar II, slated for launch in 1996, is expected to allow 250 channels of programming.

14

2

DirecTV DSS
satellite system

THE PUBLIC HAS BEEN DEMANDING A CONVENIENT AND affordable home entertainment service for many years. In this chapter we will introduce you to DirecTV,[1] the North American premiere high-powered, direct broadcast satellite (DBS) TV service. These digital signals are delivered directly to 18-inch satellite dishes installed at the home across the USA, and offers consumers up to 150 channels of quality entertainment and information programming. And with this unprecedented program variety comes digital-quality pictures and CD sound. DirecTV has used the latest advances in consumer electronics, digital compression, and satellite technologies to create an entertainment service with the flexibility, price, and quality service that TV viewers have wanted.

In the 1980s, the FCC authorized GM Hughes Electronics (GMHE) to use its considerable technological expertise and resources to develop the first home video entertainment application for direct broadcast satellite (DBS) technology. GMHE's subsidary, DirecTV, was created and went immediately to work. Combining vision with entertainment and management skill, DirecTV assembled a technical team to design and build the infrastructure required to support the first DBS system. And now the rest is history.

The Castle Rock facility

DirecTV's Castle Rock Broadcast Center is North America's first all-digital broadcast facility and the hub of the DirecTV entertainment service. The Castle Rock Center shown in Fig. 2-1 is recognized as the most sophisticated television transmission facility ever built. As the heart of DirecTV, North America's premier direct broadcast satellite service, the 55,000 square foot Castle Rock Broadcast Center (CRBC) will use the most advanced, state-of-the-art all-serial

1 Information in this chapter is courtesy of GM Hughes and DirecTV.

■ **2-1** *The DirecTV Broadcast Center, located in Castle Rock, Colorado. This satellite installation is recognized as the most sophisticated TV transmission facility ever built.*

digital equipment to uplink up to 150 channels of movies, sporting events, popular subscription networks and special attractions to two Hughes-built high power satellites. The program signals are then re-transmitted to the RCA-brand DSS (Digital Satellite System) featuring an 18-inch satellite dish installed in homes across the country.

The Broadcast Operations Control (BOC) area of the DirecTV Center (note Fig. 2-2) is where on-air programming is monitored 24 hours a day to ensure viewer satisfaction. By using the new serial digital technology and the latest state-of-the-art automation control systems throughout the CRBC, the BOC can monitor over 150 program channels. As an all-digital television transmission facility, the CRBC is very sophisticated and is the hub of DirecTV. Sony was selected for the television equipment, News Datacom for program access and encryption systems, and MEMEX Software for scheduling and program indexing, thus making this broadcast center the most sophisticated automated broadcast center ever built.

The DirecTV programming is delivered to homes across North America by two HS-601 body-stabilized satellites (see Fig. 2-3) collocated at 101 degrees West longitude. Built by Hughes Aircraft Company, the satellites measure 86 feet from wing-tip to wing-tip. Each satellite carries 16 120-watt transponders, and are the most powerful commercial satellites ever built by Hughes. Since mid-

DirecTV, Inc.

■ **2-2** *Photo of the Broadcast Operations Control (BOC) area of DirecTV in Castle Rock, CO.*

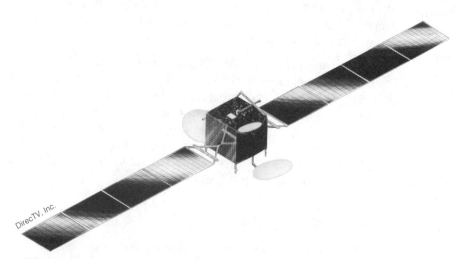

DirecTV, Inc.

■ **2-3** *The high-power HS-601 body-stabilized satellite used for the DSS system; built by GM Hughes Aircraft Company.*

1994, DirecTV has provided consumers with more than 150 channels of entertainment and information programming.

The program information is transmitted to the two high-powered birds and beamed to TV viewers, or transferred onto Betacam videotape and analyzed for the best possible viewing quality for re-broadcast later. When the tapes are declared "ready-for-air," they are digitally compressed, encrypted for security, transmitted to the satellites, and beamed to DirecTV customers.

High-power satellites

Two high-power satellites—DBS-1 and DBS-2—receive the programming signals transmitted from the Castle Rock Broadcast Center. Built by GMHE, each satellite contains 16 120-watt transponders and is in orbit 22,300 miles out from earth. Both satellites are HS-601 body-stabilized models and transmit superior signal reception, allowing DirecTV to deliver the sharpest, most reliable picture and sound quality to homes equipped with the innovative 18-inch dish. DirecTV owns 27 of the 32 transponders aboard the two satellites. Five transponders on DBS-1 are owned by United States Satellite Broadcasting (USSB).

The DSS (Digital Satellite System) receives DirecTV programming at the consumer's homes. Manufactured initially by Thomson Consumer Electronics under the RCA brand name with a suggested retail price of $699, the DSS features an 18-inch satellite dish, digital receiver, and remote control. Small and lightweight, the DSS satellite dish can be conveniently installed outside the home in locations such as a windowsill, roof line, porch railing, or a post set in concrete.

Once the system is installed and service is activated by the Castle Rock Broadcast Center, the DirecTV programming excitement begins. The DSS satellite dish captures the DirecTV programming signals, the digital receiver translates them into quality entertainment programming channels, and the remote allows viewers to scroll through an on-screen program guide and instantly select the desired program from a broad array of cable channels, pay-per view movies, sports, and much more.

Customer service support

DirecTV offers its customers the quality customer service and support subscription TV viewers have been wanting. The national customer service center, managed under contract by Matrixx Marketing, is available for DirecTV customers seven days a week, 24 hours a day.

Friendly, well-trained customer service representatives will offer multilingual support and be capable of handling up to 100,000 calls

18

a day. Whatever the type of questions, the answer is only a toll-free call away.

The national billing system—developed, operated, and managed for DirecTV by Digital Equipment Corp. and DBS Systems—allows DirecTV customers to receive their monthly programming bill in the mail, or have charges billed directly to a credit card. The system can accommodate more than 10 million customers.

DirecTV programming

Monthly subscription packages include ESPN, CNN, USA, The Disney Channel, and other major cable networks, and are a competitively-priced value compared with local cable or other satellite system subscription costs. DirecTV offers two different entertainment packages that give viewers the value-added power and flexibility to customize their own program lineup to suit individual viewing preferences. And, of course, in areas unserved by broadcast TV, network affiliate channels are available.

Direct ticket delivers more pay-per-view movies, sports, and live special events than any other cable or broadcast source. All movies are priced at $2.99 and are available long before they appear on premium cable channels. With box office hits playing as often as every 30 minutes—and no late return and rewind fees to worry about—Direct Ticket provides video store convenience right in the consumer's TV viewing rooms.

SPORTS choices include all programming shown on ESPN, TNT, USA, and Superstation TBS, plus additional professional and collegiate sporting events unavailable on network or cable channels.

A La Carte Services, such as Playboy TV and PrimeTime 24, offer DirecTV subscribers the opportunity to personalize their TV viewing with unique program choices they can pay for individually—nightly, monthly, or annually.

The DirecTV network

Over the past four years, DirecTV has invested more than half a billion dollars to assemble and manage a team of companies that is responsible for manufacturing the DSS; providing state-of-the-art digital video production and satellite transmission equipment at the Castle Rock Broadcast Center; establishing and operating a national billing system; and providing 24-hour customer service. DSS equipment and DirecTV programming are distributed by regional satellite equipment and programming distributors. DSS equipment and DirecTV programming are available to consumers through national and independent consumer electronics stores, satellite retailers, and, in rural areas, affiliates of the National Rural Telecommunications Cooperative (NRTC).

The DSS program

The DSS Digital Satellite System is forecasted to be the largest first year introduction in consumer electronics history. That means 10 to 20 million households will have DSS by the end of the decade, with sales well over $20 billion in revenues. Refer to Fig. 2-4 for these projections.

The DSS Digital Satellite System is forecasted to be the largest first year introduction in consumer electronics history. That means 10 to 12 million DSS households by the end of this decade, with sales worth over $20 billion in revenues.

Projected volume by the year 2000

	# of Households	Expected Penetration Level	Comments
No Cable Access	10–12 million	15–20% 2 million	Easiest market to penetrate. Small vs. big dish appeal. Video starved segment.
Non-subscriber with cable access	25–26 million	5–10% 2–3 million	Market unimpressed with cable programming or finds it too expensive. This segment is not video hungry or video starved.
Cable subscriber	60–62 million	10–15% 6–7 million	Supplement cable, pay-per-view appeal, out-of-market sports appeal. Programming cost is key. Video hungry.
TOTALS	**100 million**	**10–12 million**	

■ **2-4** *Projections of DSS Digital Satellite System used by the year 2000.* Thomson Consumer Electronics

DirecTV information and overview

DirecTV will deliver up to 150 channels of digital entertainment and information programming to homes equipped with low-cost, 18-inch satellite dishes. This new entertainment service gives consumers unprecedented programming choices with digital-quality picture and sound at low cost. DirecTV began service in select areas of the U.S. beginning in the summer of 1994, with full national availability by the end of 1994. DirecTV, Inc., a unit of GM Hughes

Electronics, is authorized by the Federal Communications Commission (FCC) to provide high-power direct broadcast satellite (DBS) service in the continental United States through 27 transponders at the 101 degree West longitude position.

The satellite system

DirecTV programming is distributed by two high-power HS-601 satellites built by Hughes Electronics. The first satellite (DBS-1) was launched by an Ariane rocket on December 17, 1993, and the second satellite (DBS-2) was launched on August 3, 1994, by a Martin Marietta Atlas IIA booster. Each satellite features 16 120-watt Ku-band transponders. DBS-1 delivers up to 60 channels of DirecTV programming, and 20 channels of programming from United States Satellite Broadcasting (USSB), another licensed provider. DBS-2 is used exclusively by DirecTV to bring the service up to its full capacity of approximately 150 channels. DirecTV programming is transmitted to the satellite from the DirecTV Castle Rock Broadcast Center in Castle Rock, Colorado.

The programming

DirecTV offers the broadest array of programming available to the viewing public today. Major cable TV networks comprise more than 40 of the 150 channels, including CNN, ESPN, USA Network and the Disney Channel. Another 30 channels deliver professional and collegiate sports through regional sports networks and out-of-market professional sports packages. Up to 20 channels of special interest programs are offered on an "a la carte" basis. DirecTV customers may choose from a selection of programming packages, each providing access to DirecTV Ticket, or they may purchase access to Direct Ticket only.

More DSS features

The DSS equipment features an On-Screen Program Guide, which allows customers to scan available programming and purchase pay-per-view movies with the remote control, and a Menu System that enables customers to restrict access to certain channels, build favorite-channel scan lists, and set rating/spending limits for individual movie purchases. DSS equipment is designed to the MPEG-2 digital compression standard, which ensures compatibility with new television formats such as wide-screen (16 × 9) and high-definition television (HDTV).

DSS sales and installation

The DSS equipment and DirecTV programming will be sold by national consumer electronics stores, including Sears, Lowe's, Circuit City, Best Buy, and other selected satellite retailers.

Distribution network

DirecTV has committed more than $700 million to develop the technologies and launch the national distribution infrastructure for DBS. DirecTV has assembled a team of companies that is responsible for manufacturing DSS equipment; providing state-of-the-art digital video production and satellite transmission equipment at the Castle Rock Broadcast Center; establishing and operating a national billing system; and providing 24-hour customer service.

☐ Thomson Consumer Electronics—manufactures DSS equipment; developed digital compression technology.

☐ News Datacom—developed and manages the conditional access and signal encryption systems.

☐ Sony—provided and installed broadcast equipment at the Castle Rock Broadcast Center; second manufacturer of DSS equipment.

☐ MEMEX Software—developed and manages the automated scheduling system and program library index for the Castle Rock Broadcast Center.

☐ Digital Equipment Corp.—operates and manages the national billing center.

☐ DBS Systems Corp.—provided the advanced billing software system.

☐ Matrixx Marketing—operates the DirecTV Customer Service Center and conducts telephone marketing activity seven days a week, 24 hours a day.

How the DSS technology works

Three things to remember about DSS: It is powerful, it is digital, and it is high-quality video.

If you have taken a drive across the countryside, you are aware of the TVRO TV satellite dishes (that are huge) which receive the direct broadcast signals. These TV signals have been beamed down for years to homes that look like missile tracking sites. These are the low-powered, 16-watt C-band satellites delivering an analog signal; the higher 120-watt power of the digital signal coming from each transponder of the Hughes Ku high-power satellite is what allows it to be received by such a small dish. The calculus is simple and compelling: low-power/big-dish versus high-power/small-dish. To get a handle on which delivers the better signal, just consider the difference between the light given off by a 15-watt bulb and that from a 120-watt bulb, and you will see the difference.

The signal for DSS is digital, just as the Information Superhighway is digital; data carried in ones and zeros is what is necessary for the convergence of computer data, video, and audio (in just one word, multimedia). DSS will be the first to deliver digital multimedia entertainment to a market of substantial mass.

Finally, quality is the watchword for the DSS signal. CD-quality audio and laserdisc-quality video (about 400+ lines of resolution after the conversion to MPEG-2) is higher than any previous broadcast medium.

RCA will manufacture the three-piece DSS package of equipment TV viewers will need to receive the DirecTV broadcast, which is comprised of the 18-inch satellite dish, an "integrated receiver/decoder" (IRD), and the remote control. The package is available in two configurations, both using the same dish. The basic model DS1120RW ($699) is a fully featured unit, including a 30-button remote, two A/V input/output jacks, S-video output and a wideband RS-232 data port. The step-up DS2430RW model adds the option of hooking up a second television, a 39-button universal remote, two gold-plated video inputs and four (two pairs) of R/L audio outputs. And, of course, other electronic manufacturers are interested in the DSS system. Within 18 months or with one million receivers sold by RCA, Sony will begin delivering DSS equipment.

At 18 inches in diameter and weighing only 10 pounds, the satellite receiver dish is amazingly compact. RCA recommends that professional installation be arranged by your dealer, for a cost of about $150 to $200, although a do-it-yourself kit is available for about $70. Significantly different from C-band satellites, DSS needs to be mounted only once and aimed at the satellite's stationary position (a job the DSS system itself will help you accomplish after telling it your zip code). It doesn't need to be re-aimed every time you change channels or the satellite wanders across the sky in the middle of the show you are watching. After that, save for a monsoon rain (which could temporarily affect the signal), digital programming will beam down from the satellite to your DSS dish, and then flow to the IRD in your home via copper coaxial cable. A drawing of the total DSS system is shown in (Fig. 2-5).

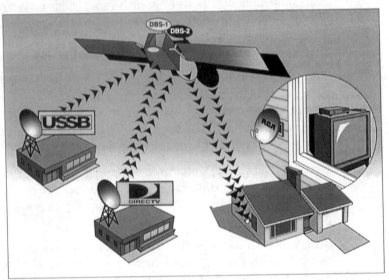

Digital programming will beam down from the satellite to your DSS dish, and then flow to the IRD in your home via copper coaxial cable.

■ **2-5** *The complete DirecTV DSS digital system.*

While the dish will be new and fun to most people, the real brains (and expense) of this Hi-Tech system is in the IRD—a black box containing the reception technology and the latest MPEG-2 digital decompression/decoding chips.

Compression and decompression are the heart of what makes DSS—and the entire Information Superhighway—possible. It is pretty hard for most people to understand how thousands of pages of text can be held on a floppy disk, but then consider that it would require a dozen or more to hold one full-size color image. And then it requires 30 of these images per second to have full-motion video. Thus, it is easy to see it would be impossible (without digital compression) to get that many ones and zeros fed into your TV set on a real-time format.

The challenge of compression is twofold. First, the program must be "encoded," or compressed down to a fraction of its full digital size, before it is sent up to the satellite; then it must be "decoded," or expanded back to its full size, when you view it. After encoding, the MPEG chips in the IRD decode, or reconstitute, the compressed signal right in the DSS receiver.

The first widely adopted compression hardware/software standard was MPEG-1 (MPEG stands for Motion Picture Experts Group). While it was a breakthrough, it just was not robust enough to handle the demands of viewers who rightly see VHS video recording quality as inadequate. MPEG encoding—the digital compression phase—works in two ways: First, after capturing the first image of a scene, it eliminates data redundancy, storing only that portion of the video image that changes from frame to frame (for example, the foreground action may change quickly while the background remains generally static). It eliminates "nonessential" image data, resulting in a degradation of image quality. The notion of what is "nonessential" has been dictated more by technological constraints than by aesthetic vision.

MPEG-1 is roughly VHS-quality in terms of horizontal resolution (about 220 to 240 lines of horizontal resolution), and tends to create "digital artifacts." While the slight herky-jerkiness of movement evident in earlier iterations of MPEG-1 has been largely remedied, it was necessary to move beyond incremental changes to a whole new system. The successor, MPEG-2, is now ready for online transmission. MPEG-2 is much smoother, and has laserdisc-level resolution (about 400 lines). RCA is the first high-volume production user of these new chips. All DSS packages sold will include MPEG-2 chips. "Service, however, will start with MPEG-1 and be upgraded to MPEG-2 once that technology is available. This transition, when it takes place, will be immediately apparent to the consumer," says James Meyer, senior vice president of Thomson Consumer Electronics. Working out the encoding for the MPEG-2 transmission will take some more time.

DSS is also an interactive system, although it is currently configured around the present-day reality that 99% of interactive traffic is what you order to come to your home (known as downstream communications), or the flow from the satellite to video viewer. But for upstream communication—to order pay-per-view programming, for example, or for the system to be able to monitor your usage and bill you accordingly—the IRD must be connected to a telephone jack in your home through its built-in modem. Any current telephone is capable of carrying the upstream data. If a telephone jack is not located near the receiver/TV, a neat solution developed by RCA is a two-unit wireless phone jack system, transmitting an FM signal (which is not affected by walls) which plugs into a standard electrical outlet near the TV set on one end and to a receiver station near an existing telephone jack at the other. The unit costs about $100.

The system tracks your regular program usage, as well as pay-per-view events and other custom programming features you may have ordered, through its SmartCard technology. SmartCard is a conditional access technology provided to DSS by News Datacom for tracking subscriber service requests and providing authorizations for subscription, advertiser-supported, and pay-per-view events. The digital encryption protection ensures program protection and prevents signal theft. You will, of course, be able to record DSS programming on your own VCR for personal viewing as you now do from local TV stations.

The national billing system—developed, operated, and managed by Digital Equipment Corp.—is capable of handling 10 million customers, and will send out monthly programming bills. If you order both DirecTV and USSB programming packages, you will receive two bills a month. The customers may also elect to have their monthly charges billed directly to a Visa or MasterCharge card.

And if you have any questions, a national customer service center will be open 24 hours a day, 365 days a year. The toll-free 800 number is projected to be able to handle 100,000 calls a day—in many different languages. The DirecTV customer information number is 1-800-347-3288 and the USSB consumer support number is 1-800-204-8772.

Satellite information

The first DSS satellite was launched in December of 1993 and is called DBS-1; the second one, DBS-2, was put into orbit at the end of 1994.

Built by GM Hughes Electronics, each body-stabilized satellite contains 16 120-watt transponders.

DirecTV owns 27 of the 32 transponders aboard the two birds, while USSB owns the other five. DBS-1 will carry approximately 60 channels of DirecTV programming, and 25 to 30 channels of USSB programming. DBS-2 is reserved exclusively for DirecTV to carry an additional 80 to 90 channels. All told, DirecTV will have about 150 to 175 channels.

Satellite uplink systems

As separate companies, DirecTV and USSB will each maintain a separate broadcast center.

Located in a 20,000-square-foot complex in Oakdale, Minn., the USSB National Broadcast Center is tied into Hubbard Broadcast's Twin Cities Headquarters and fully operating production studies, including an All News Channel. Satellite links provide access to other Hubbard news, sports, and entertainment production facilities in Florida, Washington, D.C., New Mexico, and Minnesota.

DirecTV's 55,000-square foot Castle Rock, Colo. Broadcast Center is America's first all-digital broadcast facility. Programming will be provided to Castle Rock via satellite (through eight receiving stations), over fiber optic cable, and through use of digital videotape. With more than 300 Sony Digital Betacam video recorders, a digital routing system that includes 800 inputs, and 50 automated playback and recording systems, Castle Rock is capable of transmitting 216 simultaneous channels.

Video programming

The digital video programming is broader than any cable system offering, and of much higher quality—which is what distinguishes DSS from the ordinary "ho-hum" cable system. You can receive not just HBO, but five channels of HBO, hit pay-per-view movies starting every 30 minutes, lots of audio programming, and more in the future. Thus, you make a giant leap toward your own personalized TV.

DirecTV will provide an extensive array of cable programming, including CNN, The Disney Channel, E!, The Learning Channel, USA Network, The Discovery Channel, and the (TNN) The Nashville Network. Fifty channels of pay-per-view movies from Universal Pictures, Paramount Pictures, Columbia Pictures, Tri-Star Pictures and more, with hit films available as often as every 30 minutes. Sports programming from ESPN, as well as extensive pay-per-view sporting events, should fill the needs of most various sports enthusiasts.

And with the advent of DBS-2, it will allow DirecTV to deliver Music Choice: 30 channels of CD-quality formatted music channels. The IRD has R/L audio out connections which can be connected to your A/V receiver or stereo receiver.

USSB will provide programming from HBO, Cinemax, Showtime, MTV, Nickelodeon, Lifetime, and others, an all-news channel, and free advertiser-supported channels available to anyone with a DSS system, for a total of 22 channels at the inception of the system. DirecTV and USSB have cleverly divided up the current offerings typically available on cable, with no competitive overlap between the two services.

Information overload is a big problem with the new high-tech information age. So let's look at how you can quickly sort out the 150+ channels to the one you want to watch. The on-screen menu system accesses a program guide somewhat like the TV listings found in your newspaper. But you can also view it by category, such as sports or movies, and then onto a subcategory as with golf, soccer, baseball, football, etc. Pay-per-view and special event programming is also available through the menu system.

And on into the future

The DSS may well be the start onto the electronics Super highway via digital satellite. The DSS IRD comes equipped with a modem and an RS-232 dataport, which is a computer port ready to attach to other communication services. According to Joe Clayton, senior vice president of RCA parent Thomson Consumer Electronics, "Now that the first generation of DSS is finished, you can believe that RCA engineers are busy implementing the ideas they gather into the next generation of DSS." Because DSS is fully digital, it will be forward-compatible, ready to take advantage of emerging interactive, widescreen, and HDTV video services.

Thomson, through the RCA Sarnoff Labs, is a leading contributor to the Grand Alliance, the consortium of companies defining the digital HDTV standard for the U.S. that will most likely become the world standard. Thus, DSS looks like it is well on its way for future video innovations. Right now DSS would be a great addition to a big-screen home theater with surround sound. The photo in (Fig. 2-6) shows a technician installing a DSS small dish.

The success of DSS has no doubt surprised some critics and cable TV folks, because it directly competes with cable, which supplies their own in-home equipment. However, the monthly charges are less for DSS. It was, of course, expected to be a "barn burner" in country areas with no cable hookups; but in lots of places with cable, the cable subscribers are using the DSS as well, because of the clarity of the digitally transmitted picture image and not having cable outages and other problems with cable TV companies. It also appears its causing the cable TV companies to "clean-up" their act.

The DSS system offers the same programming as cable, and has a much higher channel capacity than many cable TV systems. It also offers video-on-demand service for movies, as well as 28 channels of

■ **2-6** *A technician installing an 18-inch DSS satellite dish.*

digital audio. One disadvantage is that DSS does carry the national feeds for ABC, CBS, NBC, and FOX, it does not carry the local TV stations or independent TV stations. Thus, some customers will need outside antennas or basic cable TV. The photo in (Fig. 2-7) compares the DSS 18-inch dish with the old, analog big-dish satel-lite of yesteryear.

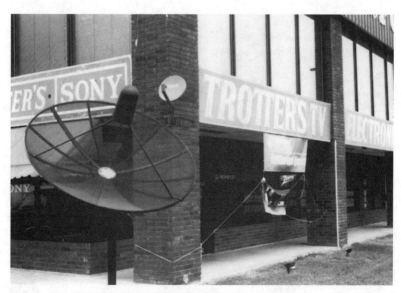

■ **2-7** *A comparison of the tiny DSS dish and the giant analog dish of yesteryear.*

And on into the future

The DSS system is currently receiving 175 channels from two satellites. A third high-power bird was launched in late 1995. As noted previously, the system is being upgraded to MPEG-2 specifications with Dolby AC-3 audio this year. When Sony starts to build units, Thomson is expected to bring out new models—under GE, ProScan, and RCA brand names—probably even a TV set with a built-in DSS unit.

Cable TV companies have also offered their own satellite system. These started in areas with no cable TV service. The PrimeStar system is now nationwide, using medium-power satellites and 39-inch satellite dishes. More competition is sure to come.

The dawning of a new video entertainment era

3

A NEW ERA IN ENTERTAINMENT BEGAN IN EARLY 1994 WITH the launch of North America's first high-power direct broadcast satellite (DBS) system. This new digital entertainment delivery system is designed to offer consumers the ultimate in programming choice, picture and sound quality, and operating convenience. Video viewers now have an affordable alternative to broadcast, cable, and videocassette viewing.

Thomson Consumer Electronics has teamed with GM Hughes Electronics to develop this revolutionary new system. GM Hughes Electronics is building and launching two advanced satellites to beam up to 150 channels of quality entertainment directly to homes equipped with Thomson's revolutionary RCA DSS high-power satellite receiving equipment. Thomson Consumer Electronics is developing the RCA brand home receiver technology featuring a revolutionary 18-inch dish antenna, and a satellite receiver capable of delivering digital laserdisc-quality video and CD-quality sound.

GM Hughes and DirecTV will provide a wide variety of programming options, including the leading subscription cable networks, hit movies from major Hollywood studies, sports and pay-per-view special events, as well as educational and cultural programming. The digital age of in-home entertainment is now a reality.

☐ The RCA brand DSS Digital Satellite System is North America's first high-powered direct broadcast satellite (DBS) system.

☐ GM Hughes Electronics is building and launching two high-power HS-601 satellites.

☐ Thomson Consumer Electronics is manufacturing the digital home entertainment technology.

- ☐ The RCA DSS Digital Satellite System consists of an 18-inch dish antenna, a satellite receiver, and a remote control.
- ☐ DSS provides the ultimate in programming choice, picture and sound quality, with:
  - Up to 150 channels of programming
  - Laserdisc-quality video capability
  - CD-quality sound capability
- ☐ Programming is to be provided by DirecTV, a unit of GM Hughes Electronics, and United States Satellite Broadcasting (USSB), a unit of Hubbard Broadcasting.
- ☐ Programming will include leading cable networks, hit movies, sports, pay-per-view events, and educational and cultural programming.
- ☐ The digital age of in-home entertainment has arrived.

Information in this chapter is courtesy of Thomson Consumer Electronics, DirecTV, and GM Hughes Electronics.

DSS market opportunity

No doubt about it, DSS is changing the way entertainment is being delivered to the American home. Not only is RCA DSS changing the home entertainment experience, but it is changing the market for direct home entertainment. Figure 3-1 shows the results from an independent market demand research study. Figure 3-2 shows the study of how the average consumer likes DSS.

Results from Independent Market Demand Research Studies

DSS Consumer Purchase Intent	Study A	Study B	Study C	Study D
Definite	6%	3%	8%	7%
Probable	16%	15%	16%	18%
TOTALS	22%	18%	24%	25%

- Overall purchase intent is extremely high.
- 8–12 million households fall into high purchase intent category.

■ 3-1 *DSS market demand research studies.* Thomson Consumer Electronics

Introduction to digital satellite system operation

Most program signals are delivered by satellite to local cable companies, who retransmit the signals to the home. This may

Major Consumer Likes Driving High Purchase Intent

Programming Options	35%
— Variety/Selection	21%
— Customization	8%
— Pay-per-view Variety	5%
18" Dish Size	20%
Suggested Retail Price Point	17%
Ease of Installation	6%

■ **3-2** *DSS market opportunity percentages.* Thomson Consumer Electronics

cause significant signal loss and, in turn, lost picture quality. Now viewers can have crystal-clear signals transmitted directly to their home.

It all starts at DirecTV and USSB, the national television programming companies who provide up to 150 channels of entertainment and information. DirecTV and USSB initiate the programming from two uplink centers, where program information is digitized, compressed, and beamed to two Hughes HS-601 satellites traveling in orbit 22,300 miles above the equator. These powerful satellites receive the digital programming information, then relay it directly to homes anywhere in the continental United States.

Because the Hughes satellite transponders are so powerful, the RCA DSS dish only needs to be 18 inches in diameter. And because the satellites are in a fixed orbit, homeowners aim the dish once and forget it. When the signal reaches the RCA DSS dish, it travels to the satellite receiver inside the consumer's home.

Unlike some ordinary broadcast transmission, digital signals reach consumers homes ghost-free and capable of delivering a laserdisc-quality picture and CD-quality sound. And the RCA DSS satellite receiver is linked to the DirecTV and USSB Customer Service Centers through a phone jack on the back of the satellite receiver. This telephone hookup makes ordering pay-per-view events as convenient as pushing a button on the remote control.

Yet, with all this space-age digital technology, RCA DSS is easy to use. One simple remote control operates both the RCA DSS satellite receiver and most brands of televisions. Just point, select, and enjoy. It's just that simple.

Summary of how DSS works

- [ ] DSS satellites transmit crystal-clear digital signals directly to your home.
- [ ] DirecTV initiates programming from their uplink center in Castle Rock, Colorado.
- [ ] USSB initiates programming from their uplink center in Oakdale, Minnesota.
- [ ] Program information is digitized and compressed, then beamed to two Hughes HS-601 satellites.
- [ ] The satellites receive the digital programming information, then relay it back to consumer's homes.
- [ ] The 18-inch RCA DSS satellite dish antenna receives the signal.
- [ ] The RCA satellite receiver decompresses and decodes the signal for consumer viewing.
- [ ] The consumer or installer simply aims the dish once, then forgets it.
- [ ] The RCA DSS satellite receiver is linked to the DirecTV and USSB Customer Service Centers through a phone jack; the telephone hookup is used for billing.
- [ ] The remote control operates both the RCA DSS satellite receiver and most brands of televisions.

Brief DSS user operation information

The RCA DSS Digital Satellite System is advanced digital technology—yet it is easy to use as an ordinary TV set. To get started, a viewer simply picks up the remote control and presses the "ON" button.

Using the remote, viewers can scan through all the listings, select a favorite show, or tell the Program Guide to list specific types of programs: sports, movies, children's programming, and so on. Choices are made by using the arrow keys on the remote control or the satellite receiver. The arrows let the viewer point to an item, and then press the "Select" key to choose that program.

If the viewer selects a program that is to be aired in the future, program details will appear. If the viewer chooses a pay-per-view event, a screen prompt allows the viewer to purchase the program, view information about the program, or return to the Program Guide.

Features such as Program Ratings and Program Spending Limits help consumers manage their viewing time and options. To help consumers operate all of the features of the RCA DSS satellite receiver, an On-Screen Menu System is provided. Viewers can use the menu to help setup the system, build channel lists, preview coming attractions, order pay-per-view selections, set viewing limits, obtain help, and receive mail from the DirecTV and USSB program providers.

There are two parts to the Menu System: menus, and display screens. A menu is simply a list of choices. As viewers use the menu system, "help" messages will appear to explain what is happening or to prompt them on what to do next. Display screens are the working screens that are generated from menu selections. Simply point and select the option desired.

DSS operation summary

Let's now look at a summary of what makes the DSS system so easy to use.

The DSS user interface

The following items are provided by the on-screen menu interface:
- ☐ Ratings and spending limits
- ☐ Favorite channel lists
- ☐ Coming attractions of pay-per-view events
- ☐ Alternate audio programs
- ☐ An online help system
- ☐ An online mail system
- ☐ An online system test for quick diagnostics

The DSS program guide

DSS program guide features:
- ☐ User-friendly "point and select" operation.
- ☐ Lists of available programs and times.
- ☐ Detailed program information.
- ☐ It sorts programs by categories, such as sports or movies.
- ☐ It sorts programs by consumers favorite channel lists.

Figure 3-3, on the next page, shows how the screen menu is set up.

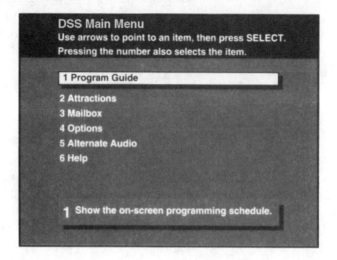

■ **3-3** *DSS user menu screen formats.*

The DSS receiver

The DSS receiver has a built-in switch so that a TV antenna or cable can be connected into the receiver. Because of FCC regulations, some network signals cannot be sent via DSS to areas where they can be received by a local TV station.

The rear panel of the receiver has a modular phone jack. The phone connection is required so that monthly charges from pay-per-view and other specials can be reported to the programming providers. It also keeps tabs on the receiver in case of theft or the receiver's being moved from its assigned address; it must not be changed, because of regional program distribution.

The DSS user interface is an on-screen point-and-shoot menu. The system also allows for the review of current billing information. The DSS receiver also has other programmable features. The receiver allows pay-per-view locks and limits that can be set. Parents could, as an example, program per-event spending limits or rating ceilings to prevent their children from accessing undesirable programs. The entire system can also be set to be locked unless the proper four-digit code is entered.

The front of the receiver has a slot for the "smart card" insertion. The smart card contains subscription and decoding information and records programs that you wish to view. The conditional-access encryption method was developed by News Datacom.

The RCA executives like to stress their ability to shift from MPEG-1 to MPEG-2 as an example of the forward/compatibility of the DSS system. RCA states that the DSS system has the capability for receiving 16×9 wide-screen format TV signals (as well as HDTV broadcasts once that revolutionary technology is in place), which means no "interim technology" with the introduction of RCA DSS.

More Hughes satellite data

The 32 transponders on the two DBS satellites can distribute about 150 channels of programming, thanks to digital "compression" or bit rate reduction. An exact number of channels cannot be given because not all programming will be compressed the same way. The compression ratio for a sporting event would be much lower than for an old black-and-white movie. The frame rate of the source material, and the desired quality of the output, also affects the compression ratio that can be achieved. Hughes estimates that each transponder can broadcast up to four live video signals or up to eight movies simultaneously.

The amplifiers, according to Hughes, will amplify either analog or digital signals, and will be capable of transmitting HDTV signals and CD-quality audio. The uplink signals are sent from the DirecTV Castle Rock Broadcast Center in Colorado. This high-tech operation is capable of transmitting up to 216 simultaneous broadcast channels to the two satellites.

The satellites have an expected operational lifetime of about 12 years. Although the electronics on board will probably be operating much longer, the spacecraft will exhaust its fuel supply. Geostationary orbiting satellites require propellants to ensure their precise positioning, because the gravitational effects of the sun and moon can move them out of position.

Will DSS be a success?

As of this writing, the service is going great guns and appears to be a "barn burner" in the months ahead. RCA has now sold over 500,000 receiver/dish units. All of the companies involved in the DSS project are very optimistic about its chance for success. The first target will be the 10 million rural homes that have no access to cable TV. The service does not actually offer any new programs or movies; Thomson is confident that it can also attract some of the 20 to 30 million homes that have access to cable but do not subscribe. And it now appears that many of the large number of cable viewers who are dissatisfied with their cable company are installing DSS dishes.

The expected break-even point for both Hughes and USSB, the two big investors, is only about 3% of all homes, or 3 million subscribers. And DirecTV seems to think that, stressing the system's value, flexibility, and performance, they would be elated with 10% penetration over the next few years. DirecTV thinks they will achieve this if the market research is anywhere near correct.

Thomson stresses that even those homeowners who do not subscribe to cable—although they have access to it—rent videotapes, often on a daily basis. Pay-per-view programming eliminates the need to travel to a video rental store.

DSS picture quality

Only large-dish satellite systems and videodiscs can offer similar quality. DSS also has a larger channel capacity than any existing cable system, and the pricing packages presented at this time are less expensive than similar cable packages.

Some analysts point out that DSS could be superseded by the electronic information superhighway and its many projected interactive systems. DSS is here right now, however. The viewers' wants for more programming has been proven by cable (near 70%) and by video rentals (over a $4 billion annual industry). And whether consumers want to be able to interact with their TV sets is a question that is still waiting for the jury's decision.

One other "bugaboo" for the DSS system is that a TV viewer who wants to receive digital satellite programming must shell out $700 or more. However, it has been pointed out that $700 is a small portion of the cost for a large-screen TV receiver that requires a very high-quality signal for best results.

But one thing is for sure: TV viewers have another entertainment choice, and it appears that DSS will give the cable TV folks a run for their money.

DSS customer's information glossary

Access card The access card identifies the user to the DSS service providers and is required for the Digital Satellite System to operate.

Alignment, dish The process of adjusting the dish antenna in order to maximize its signal-receiving capabilities.

Alternate audio Alternate audio refers to the different audio channels that may be broadcast in conjunction with a video service. An example of this is a foreign-language translation.

Attractions Preview of special programs to be broadcast by the programming service provider.

Azimuth The number of degrees of rotation on a compass in a clockwise direction from true north. This information can be used to determine the direction where the satellite is located in relation to the home, and to help in pointing the dish antenna in a left-to-right direction towards the satellite.

Bullet amplifier A small device that is used to increase the power of the signal, in order to compensate for the losses induced by coaxial cable and signal splitting devices.

Coaxial cable A specific type of cable that is used to transmit high-frequency signals with a minimum amount of loss. RG-6 coax is the type of cable used to connect the DSS dish antenna to the satellite receiver.

Dish That portion of the satellite antenna that is used to collect, reflect, and focus the satellite signals into the LNB.

Drip loop Several inches of slack in a cable that helps prevent water from collecting on the cable and running along the surface of the cable into a connector or entry point.

DSS receiver A satellite receiver that receives, processes, and converts the digitally compressed satellite signals in picture and sound.

Earth ground A conducting connection to the earth for an electrical charge.

Elevation A vertical angle that is measured from the horizon up to the satellite. This information can be used to determine how far up in the sky the satellite is located in relation to your home, and to help in pointing the dish antenna in an up-and-down direction toward the satellite.

F connector A special screw-on connector that is crimped onto the coax cable. This F connector is commonly used with RG-59U and RG-6U coaxial cable.

Feedhorn The input to the LNB that collects the focused signals reflected by the dish.

Filtering A means of sorting and limiting the amount and type of information displayed in the program guide.

Geostationary A circular orbit approximately 22,300 miles above the equator, in which satellites travel at the same apparent speed as the rotation of the earth.

Grounding block A device that connects two coaxial cables together and is used with a ground wire to connect the cable to earth ground. This is to help prevent electrical surges from being transmitted by the coax.

Ground rod A metal rod, typically eight feet in length, that is driven into the ground in order to establish a conductive path to the earth.

Ground wire The type of wire that is used to connect the antenna and grounding block to earth ground.

Key A user-designated 4-digit password that allows access to be limited to certain capabilities of the DSS receiver.

Latitude The distance, measured in degrees, of a position on the surface of the earth that is north or south of the equator.

Limits The user defined feature that allows program viewing control by rating or on a cost-per-program basis.

LNB An acronym that stands for Low Noise Block converter. The LNB is mounted at the focal point of the dish. It is used to amplify and convert the satellite signals into a block of frequencies that can be used in conjunction with standard coaxial cable, and can be received by the tuner in the DSS receiver.

Locks The means to restrict access to certain capabilities of the DSS receiver. The lock is turned on or off by the 4-digit key. The condition of the lock is indicated by an open or closed lock icon that is displayed in the channel marker area.

Longitude The distance, measured in degrees, of a position on the surface of the earth that is east or west of the prime meridian.

Mailbox A place where incoming electronic messages that are transmitted by the service provider can be stored and retrieved. Each message contains an expiration date and will be automatically deleted if not erased prior to that date. The mailbox can store up to 400 characters and up to 10 messages (each message takes a minimum of 40 characters) as long as the total character count does not exceed 400. Once the maximum storage capacity has been reached, messages will be automatically deleted based on the system requirements established by the service providers.

Main menu A list of choices displayed by the DSS receiver's on-screen menu system when at first the menu button is pressed.

Mast A vertically aligned 1½-inch diameter pipe that is used to support the satellite dish support arm.

Mounting foot That portion of the DSS dish antenna assembly that is used to mount the mast to a supporting structure.

Pending purchases A detailed list of programs purchased but not yet viewed. Between the pending purchases and purchase history, the system can store as many as 25 transactions.

Phone prefix A single digit (such as "9") that must be dialed first in order to access an outside line before dialing the actual phone number.

Plumb line A string of sufficient length that has a weight attached to one end that, when extended, can provide a reference line perpendicular to ground.

Point and select The method of using the remote control to navigate to different parts of the on-screen display and to choose highlighted choices.

Preterminated coaxial cable RG-6 type coaxial cable with factory installed F connectors.

41

Program guide An electronic listing of programs and services available from the service providers.

Purchase history A detailed list of programs previously purchased and viewed. Between the pending purchases and purchase history, the system can store as many as 25 transactions. The purchase history is periodically downloaded to the service provider through the telephone connection, and may or may not be erased when the information is retrieved.

Rafter A timber, usually spaced 16 inches apart, that is used to support a roof.

Rating limit Provides a means of restricting viewing based on ratings of movies utilizing the Motion Picture Actors Association ratings. Operation of this feature is dependent upon the accuracy of the data being supplied and broadcast by the service provider.

Signal meter A DSS receiver-generated on-screen display that indicates the relative strength of the satellite signal and can be used to fine tune the position of the dish by adjusting the dish until the maximum value is indicated.

Spending limit This provides a means of restricting viewing based on a cost per program limit.

Stud A vertical timber, usually spaced 16 inches apart, that is used to support a wall.

System test A feature that provides some level of diagnostics and can be used to determine that the main components of the DSS system are working properly.

Themes A method of sorting the program guide based on programming categories, in order to make it easier to find a program of interest.

TV/DSS A button that is used to toggle the DSS receivers out to TV connectors, which may be connected to the television's antenna input, between antenna or cable delivered programming and DSS satellite programming. The operation is similar to the TV/VCR button switching system found on VCRs.

DSS system installation

4

THIS CHAPTER CONTAINS INFORMATION YOU CAN USE FOR DSS dish and receiver installation or as an owner of a DSS video satellite system.

Overview

The following section outlines the procedures for setting up a DSS installation.

Getting started

This section describes the contents of the DSS dish carton and instructions for assembling the dish. It provides descriptions of the materials and tools needed for dish installation.

Preinstallation activities

This section provides information about the procedures you should follow when locating the best view of the satellite from your house and determining the best site for the dish. You can use this information to decide whether you will install the dish yourself or contact a professional installer.

Mounting the dish

This section provides detailed instructions for mounting the dish on different surfaces and in different locations. It also provides instructions for vertically aligning parts of the dish so that your system will receive the strongest satellite signal.

Grounding the dish

This section gives instructions for grounding the dish, grounding a metal pole in a pole mount, and connecting the grounding lightning protection block.

Installing and routing the cable

This section gives instructions for routing the coax cable from the dish to the grounding block and from the grounding block to the

DSS receiver. This includes basic information about routing the cable through an exterior wall and an interior floor.

Pointing the satellite dish

This part contains instructions for setting the elevation of the dish, and for pointing the dish toward the satellite so that your system can receive the strongest possible satellite signal.

DSS dish carton

The DSS dish carton contains the following parts (see Fig. 4-1):

- ☐ Mounting foot assembly and mast
- ☐ Dish (or reflector)
- ☐ LNB (Low Noise Block) converter
- ☐ LNB support arm assembly
- ☐ Hardware packet

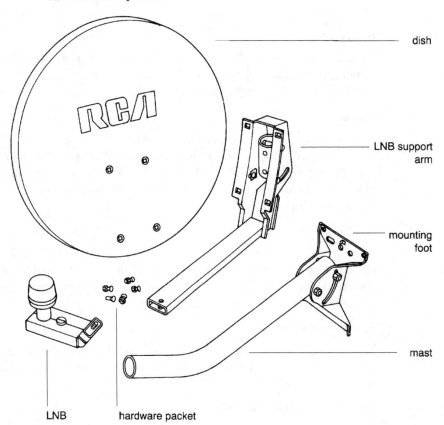

dish

LNB support arm

mounting foot

mast

LNB   hardware packet

■ **4-1** *Contents of DSS dish carton.* Thomson Consumer Electronics

Assembling the dish

The bolts you need to assemble the dish are in a plastic packet that came in the DSS dish carton. As you handle the dish parts, avoid scratching the surface of the dish. Scratches can damage the protective surface.

Most of the dish parts were preassembled for you at the factory. You will be given detailed instruction in this chapter for any required assembly. The drawing in Fig. 4-2 shows how the dish parts fit together.

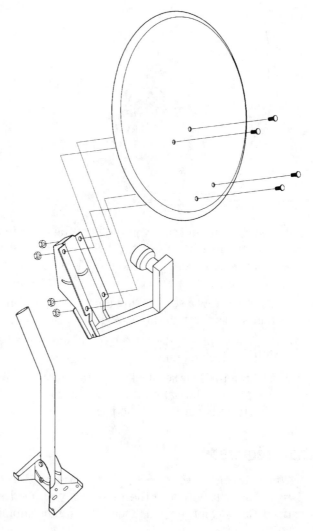

■ **4-2** *How the dish parts fit together.* Thomson Consumer Electronics

Attaching the dish to the LNB support arm

1. Locate the four flathead screws and the four self-locking nuts from the hardware packet that came with the DSS dish.

2. Press the mounting bracket of the LNB support arm to the back of the dish, matching the four square holes on the bracket to those on the dish as shown in Fig. 4-3.

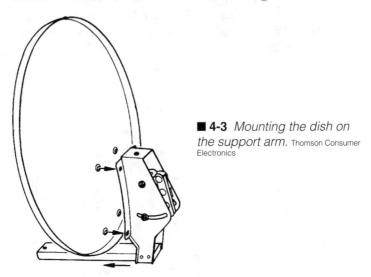

■ **4-3** *Mounting the dish on the support arm.* Thomson Consumer Electronics

3. Insert a flathead screw into one of the holes in the face of the dish and gently press it through the matching hole in the mounting bracket. The wide, flat head of the screw should be on the inside of the dish.

4. Lightly screw a self-locking nut on the portion of the screw that sticks through the back of the dish and mounting bracket.

5. Repeat steps 3 and 4 for the remaining three flathead screws and self-locking nuts.

6. Tighten the four self-locking nuts. The heads of the flathead screws must be flat on the inside surface of the dish for the dish to be securely fastened to the LNB support arm.

Materials required

When you install your DSS dish, you will need the following materials. You can purchase the materials you need from any authorized DSS dealer, or your local electronics supplier or hardware store.

Coaxial cable

You will need RG-6U coaxial cable with weatherproof "F" connectors at both ends to:

☐ Connect the LNB to the grounding block.

☐ Connect the grounding block to the DSS receiver.

If you route the coaxial cable through an interior wall and install a coaxial cable wall plate, you will need an additional piece of coaxial cable to connect the coaxial cable connector on the wall plate to the DSS receiver.

Ground wire

You will need a #8 aluminum or a #10 copper ground wire to use as follows:

☐ Connect the grounding block to a ground rod.

☐ Connect the ground bolt on the dish to a ground rod.

NOTE: Do NOT splice any ground wire run. Obtain the correct length for each application.

Grounding block

You will need a grounding block to ground the coaxial cable.

Tools needed for DSS dish installation

To install the DSS dish, you will need the following tools, as shown on the following page in Fig. 4-4.

☐ $\frac{7}{16}$" and $\frac{1}{2}$" open-end wrenches. You can use an adjustable wrench, ratchet tool with $\frac{7}{16}$" and $\frac{1}{2}$" sockets, or a hex nut tool. Do not use any type of pliers. The teeth of the pliers may damage the protective surface of the dish parts.

☐ A plumb line or bubble level. A plumb line is a piece of string at least 2 feet long, with a small lead weight, a heavy nut, or a bolt tied to one end.

☐ An electric drill and drill bits.

☐ A pencil.

☐ Screwdrivers (Phillips and blade).

☐ A compass.

47

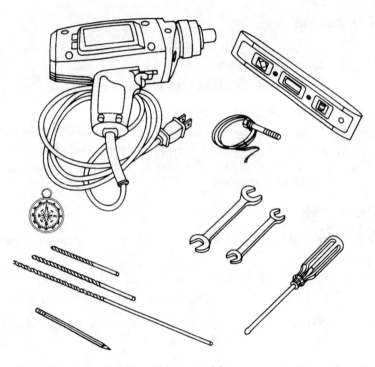

■ **4-4** *Tools required for dish assembly.* Thomson Consumer Electronics

Preinstallation information

Before you start to install your dish, plan the entire installation carefully.

1. Find the satellite in relationship to your house.
2. Identify potential mounting sites.
3. Identify potential mounting options.
4. Identify the cable requirements.
5. Select the mounting site.

Finding the satellite

The satellites are parked in a geostationary orbit over the equator, south of Texas, at the 101° west longitude position. Refer to the map in Fig. 4-5.

Potential dish mounting site considerations

You have located the satellite and know the general direction in which the dish must point. Now you can further your site selection. Refer to Fig. 4-6.

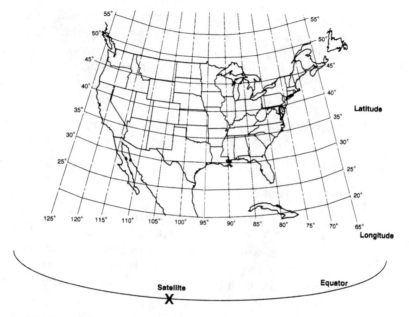

■ 4-5 *A map showing the location of the DSS satellite.* Thomson Consumer Electronics

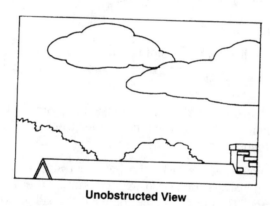

Unobstructed View

Obstructed View

Thomson Consumer Electronics

■ 4-6 *Selection of the dish mounting site.*

The southern view must be unobstructed, now and in the future. There can be no trees or structures between the DSS dish and the satellite. Thus, take into consideration seasonal foliage, future growth of existing trees, possible remodeling or additions to your house, construction of other buildings, and changes to the landscaping.

WARNING: Do not install the dish near power lines, electric lights or power circuits. It is recommended that the dish be located more than 20 feet from overhead power lines.

Various mounting options

There are several dish mounting options for you to consider. The one you choose depends on the site you selected. However, the structure on which you decide to mount the dish must be stable and secure. Even moderate winds can create several hundred pounds of force on the screws attaching the mounting foot to the structure.

When you select potential mounting sites, review the following mounting considerations before you select the final site.

Mounting the dish on wood

You can mount the dish on wood siding, on a deck, or on a roof as shown in Fig. 4-7. However, use the roof mount only as a last resort.

If you mount the dish on the side of your house, make sure you attach the dish's mounting foot to the studs under the siding. This provides a more secure mount.

Take care when mounting the dish on asbestos or stucco siding, because you can crack the siding. Also, avoid mounting the dish on aluminum or composite siding. You can damage the siding when you tighten the screws.

Use a roof mount as a last resort, because you can easily damage your roof and create leaks. If you decide to mount the DSS dish on your roof, proceed with caution. Use your ladder with care.

Roof mount considerations

☐ Make sure the roof provides an unobstructed view of the satellite.

☐ Do not mount the dish on slate or shake shingles.

■ 4-7 *How to mount the dish on wood.*

Thomson Consumer Electronics

☐ Do not mount the dish on an overhang.

☐ Mount the DSS dish over the outer wall or over a rafter.

☐ On a flat roof, do not mount the dish on a low place where water collects. Mounting the dish on a higher place helps ensure the silicone sealant effectively seals any holes you drill.

The materials needed to mount the dish on a wood surface depend on the construction of the surface. The following lists identify the materials needed for various installations.

Hollow walls

☐ Four BB-¼" togglers

☐ Four ¼-20 × 3" machine screws

☐ Four ⁵⁄₁₆" washers

Wall studs or rafters

☐ Two 3" × ⁵⁄₁₆" lag screws

☐ Two ⁵⁄₁₆" washers

Solid wood surfaces

☐ Four 2" × ⁵⁄₁₆" lag screws

☐ Four ⁵⁄₁₆" washers

Tools needed

☐ A plumb line or bubble level

☐ A screwdriver

☐ ⁷⁄₁₆" and a ⅜" open-end wrench or an adjustable, hex nut, or ratchet wrench

☐ A pencil

☐ An electric drill and a ½" or ⁷⁄₃₂" high-speed-steel bit for drilling wood

Mounting the dish on brick or cinder block

You can mount the dish on brick, poured concrete, or cinder block (note Fig. 4-8).

Thomson Consumer Electronics

■ **4-8** *A dish mounted on a brick wall.*

When you drill into brick, do not drill into the mortar between the bricks. Drilling into the mortar can damage the mortar and may cause the dish to be unstable. If you are unsure whether drilling into the mortar or brick on your house will damage either, consult a contractor.

When you drill into brick or poured concrete, you will need to drill a ½" hole at least 3 inches deep for the double expansion anchors and to accommodate the length of the machine screws. To make drilling easier, begin drilling with a smaller masonry drill bit, then switch to the ½" masonry drill bit. If your house is sided with brick veneer, make sure you locate the studs under the veneer so that you can securely attach the mounting foot to the wall.

Materials required to mount the dish on a brick, poured concrete, or cinder block surface depend on the construction surface.

The following lists name the materials needed for installation on masonry surfaces:

Brick or poured concrete

☐ Four B4015 or equivalent double-expansion anchors

☐ Four ¼-20 × 3" machine screws

Cinder block

☐ Two togglers

☐ Two ¼-20 × 3" machine screws

Tools needed

☐ A plumb line or bubble level

☐ A screwdriver

☐ A hammer

☐ A pencil

☐ An electric drill and a ½" masonry bit for drilling into brick, poured concrete, or cinder block.

Mounting the dish on a pole

You can mount the dish on a pole made from rust-resistant metal pipe. One end of the pole is embedded in concrete and buried in the ground. Before you start a pole mount, make sure that you consider future landscaping, gardens, pools, or additions to your house.

If you mount the dish on a pole beside your house, determine the height at which you want the dish. Then add three feet to accommodate the portion of the pole that will be buried below ground (see Fig. 4-9). Also, you may need to increase the length to make sure the bottom of the pole is at least 6 inches below the frost line in your area. To find out the depth of the frost line in your area, call the local agricultural extension office. Make sure the pole you purchase is rust-resistant.

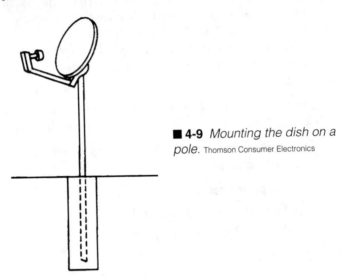

■ **4-9** *Mounting the dish on a pole*. Thomson Consumer Electronics

Also, consider the soil conditions in the mounting location. Do not install the pole in wet or marshy areas, because the cement may not cure properly and the pole may become unstable. If the ground is rocky or extremely hard, you may not be able to dig a hole that is at least 3 feet deep, the required depth.

If the length of pipe above ground is too long, the dish may be unstable in windy weather. However, you can install guy wires to increase its stability.

To prepare the metal pole (see Fig. 4-10):

☐ Drill a ¼" hole into the pole approximately one foot from the top. This hole is used for the zinc-plated ¼-20 × 1¾" bolt, ¼" star washer, and ¼" nut that attach the ground wire to the pole.

☐ Cut the end of the pole that is to be buried at an angle to prevent the pipe from moving in the cement.

■ **4-10** *Preparing the metal mounting pole.* Thomson Consumer Electronics

If you mount the dish on a pole, you will need the following materials and tools:

- ☐ A 1¼" ID Schedule 40 rust-resistant metal pipe, three feet longer than the mounting height of the dish
- ☐ A ¼-20 × 1¾" zinc-plated bolt with ¼" nut and ¼" star washer
- ☐ Quick-drying cement
- ☐ Four guy wires and stakes.

Tools needed

- ☐ A plumb line or bubble level
- ☐ A shovel or posthole digger
- ☐ A hacksaw
- ☐ An adjustable wrench

Mounting the dish on a chimney

Use the chimney mount only if you cannot locate a good mounting site on the side of your house, on your deck patio, or on a pole next to your house. High winds can put great strain on the mounting and the chimney if the dish is mounted improperly (see Fig. 4-11 on the following page).

Make sure that the chimney is sturdy and in good condition. Also, make sure the dish is properly mounted using a chimney mount kit so the dish and the chimney can withstand any strain from high

■ **4-11** *A dish mounted on a chimney.*

winds. You can purchase a chimney mount kit from your local electronics dealer.

Make sure the chimney provides an unobstructed view of the satellite location. The chimney should be tall enough that the top of the dish does not extend above the top of the chimney. This prevents damage to the dish from heat and soot.

If you choose to mount the dish on the chimney, you will need the following materials and tools:

☐ Chimney mount kit.

Tools required

☐ Plumb line or bubble level

☐ Screwdriver

☐ Adjustable wrench

Identify cable requirements

After you select potential sites and mounting options, decide where the cable from the dish should enter the house. Then, measure the distance from the dish to the DSS receiver. If possible, pick the shortest and most direct route from the dish to the DSS receiver. The examples in Fig. 4-12 illustrate the cable routed through a crawl space and through the attic.

If you need more than 112 feet of cable in your installation, you must install a TVRO bullet amplifier at the end of the coax cable

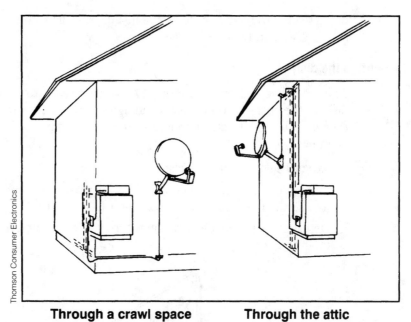

Thomson Consumer Electronics

Through a crawl space   **Through the attic**

■ **4-12** *How the cables can be installed.*

before you add additional cable length. The amplifier boosts the signal strength.

Selecting the mounting site

After you find the correct view of the satellite and locate several sites, each with an unobstructed view of the satellite, evaluate your options. Then select the site that provides the best view of the satellite and the structural integrity needed for mounting the dish. Also, evaluate your comfort in installing the dish yourself.

After you review your options, if you decide to install the dish yourself, you should feel comfortable with the types of construction practices that may be necessary for your installation. These might include:

☐ Climbing ladders

☐ Climbing on your roof

☐ Drilling holes in your house or roof

☐ Working around power lines

☐ Working with cement

☐ Routing coaxial cable

☐ Grounding the dish with a grounding block, grounding rod, and ground wire.

If you decide to have a professional install your dish, contact your local authorized DSS dealer.

Mounting the dish

After you have chosen a mounting site, you are ready to mount the dish on the surface you have chosen. Let's now see how to mount the dish on wood, brick, or concrete surfaces, on a pole, or on your chimney.

Before you begin mounting the dish

☐ Tighten one of the two bolts holding the mast to the mounting foot so the mast does not move. Refer to Fig. 4-13.

☐ Determine the specific location where you will attach the mounting foot.

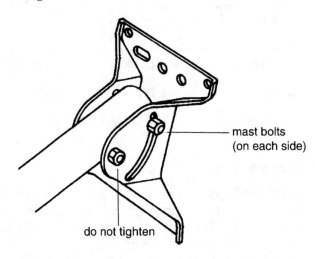

■ **4-13** *Mounting the mast to the mounting foot.* Thomson Consumer Electronics

Attaching the mounting foot to wood

Make sure you have the materials and tools you need. The materials needed to mount the dish on a wood surface depend on the construction of the surface.

☐ If the surface is vertical, such as the side of a house, use a plumb line or bubble level to make sure the mounting foot is perfectly vertical (see Fig. 4-14). Make sure you locate and secure the mounting foot to the center of a wall stud. Do not mount the dish near the edge of a stud.

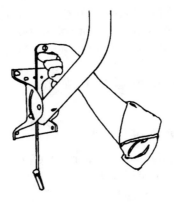

■ **4-14** *Using a plumb line to mount the foot.* Thomson Consumer Electronics

☐ If the surface is diagonal, such as a roof, you can line up the mounting foot with the shingles visually. Also, if the surface has rafters, make sure you locate and secure the mounting foot to a rafter. To attach the mounting foot, follow these steps:

1. Hold the mounting foot in position on the mounting surface and, using the mounting foot as a template, mark the appropriate holes. For wall studs or rafters, center the mounting foot on the rafter and mark the two center holes as shown in Fig. 4-15. For hollow wall and solid wood construction, mark the four outside holes on the mounting, as shown in Fig. 4-16.

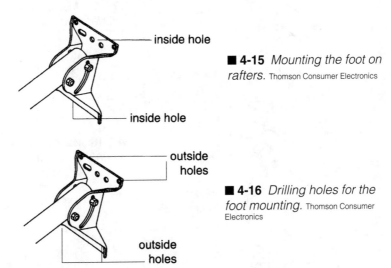

inside hole

inside hole

■ **4-15** *Mounting the foot on rafters.* Thomson Consumer Electronics

outside holes

outside holes

■ **4-16** *Drilling holes for the foot mounting.* Thomson Consumer Electronics

2. Remove the mounting foot and drill the holes in the locations you marked. For studs or solid wood, drill $\frac{7}{32}$" holes; for hollow wall construction, drill $\frac{1}{2}$" holes and insert togglers.

3. Hold the mounting foot in place, then insert and lightly tighten the lag screws for studs or solid wood, or insert machine screws with washers into the toggles for hollow walls.

4. Securely tighten the screws. You have finished attaching the mounting foot.

Mounting on brick, poured concrete, or cinder block

Make sure you have the tools and materials required for the job. When all the required elements have been assembled, follow these steps:

1. Hold the mounting foot in position and, using the mounting foot as a template, mark the holes with a pencil. If the surface is brick or poured concrete, mark the four outside holes on the mounting foot as shown in Fig. 4-17. If the surface is cinder block, mark the two center holes as shown in Fig. 4-18. If it is a vertical surface, you must use a plumb line or bubble level to make sure the mounting foot is perfectly vertical, as shown in Fig. 4-19.

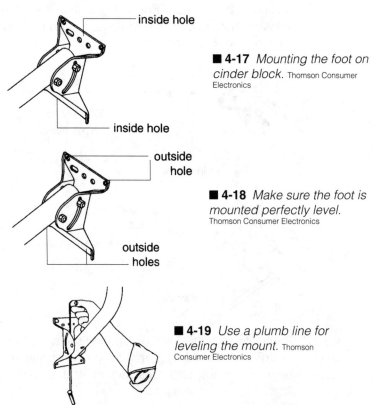

inside hole

inside hole

■ **4-17** *Mounting the foot on cinder block.* Thomson Consumer Electronics

outside hole

outside holes

■ **4-18** *Make sure the foot is mounted perfectly level.* Thomson Consumer Electronics

■ **4-19** *Use a plumb line for leveling the mount.* Thomson Consumer Electronics

2. Remove the mounting foot. Then drill four holes if the surface is solid, or two holes if the surface is cinder block.

3. Insert four B4015 or equivalent double-expansion anchors for a solid surface, or two togglers for cinder block.

4. Place the mounting foot on the surface, matching the holes on the mounting surface.

5. Insert and tighten the machine screws. You have now finished attaching the mounting foot.

Mounting on a metal pole

Before you start, make sure you have the materials and tools you need. Please note that the outer diameter (OD) of the pole must be approximately 1⅝". This ensures that the dish-mounting assembly will fit on top of the pole, and that you will be able to tighten the mounting assembly when you have correctly pointed the dish. The recommended pipe (1¼" ID Schedule 40 water pipe) provides the correct outer diameter.

The bottom of the pole must be at least 6" below the frost line. To find out the frost line in your area, call the local agricultural extension service. Remember, you will have to reach to the top of the pole to install the dish, so make sure you will be able to reach it.

Prepare to install the pole

1. Dig a 12" diameter hole 3 feet deep, in the location where you want the pole.

2. Place the metal pole in the hole and backfill the hole with just enough dirt to hold the pole upright.

3. Attach four temporary guy wires to the top of the pole (see Fig. 4-20 on the following page).

4. Make sure the pole is perfectly vertical in all directions. To do this, hold the plumb line or bubble level next to the pole, as shown in Fig. 4-21 on the following page. Then gently move the pole so that it aligns with the leveling tool you use.

5. Tighten the guy wires to hold the pole in a vertical position until the cement is dry.

6. Fill the hole to within two inches of ground level with prepared, quick-drying cement, and let the cement dry completely before you remove the guy wires. You have finished installing the pole, and the LNB support arm can now be attached.

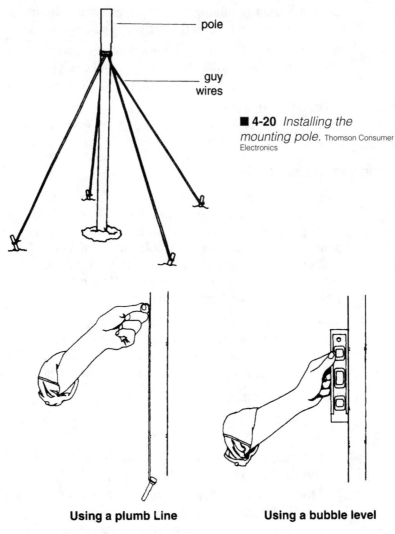

pole

guy
wires

■ **4-20** *Installing the
mounting pole.* Thomson Consumer
Electronics

Using a plumb Line   **Using a bubble level**

■ **4-21** *Using a plumb line or a bubble level.* Thomson Consumer Electronics

Mounting the dish on a chimney

Use the chimney mount only if you cannot use another location.
Also, mount the dish on your chimney only if the chimney is
sturdy. Chimneys that are old or not structurally sound can be
damaged easily by dish vibrations caused by even moderate winds.

Before you start, make sure you have the materials and tools you
need. If you purchase a chimney mount kit from an electronics supplier or hardware store, install the kit according to the instructions

with the kit. When you install the rust-resistant metal pipe used as the mast in most chimney mount kits, vertically align the pipe.

Installing the DSS chimney mount kit

1. Each metal strap in the chimney mount kit has a preattached eyebolt. Insert an eyebolt in the appropriate hole in the chimney bracket, as shown in Fig. 4-22.

■ **4-22** *Installing the eyebolt.*
Thomson Consumer Electronics

2. Place a nut onto the eyebolt and tighten it to about ½ inch from the end of the eyebolt.
3. Repeat steps 1 and 2 for the second metal strap.
4. Press the chimney bracket on the corner of the chimney as shown in Fig. 4-23.

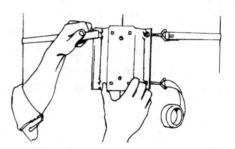

■ **4-23** *Installing the chimney brackets.*
Thomson Consumer Electronics

5. Stretch one metal strap around the chimney. Do not twist the strap.
6. Pull the strap tight around the chimney. Make sure the strap is horizontal all the way around.
7. Insert the free end of the strap through the eyebolt supplied with the kit. Then, attach the strap clamps as shown in Fig. 4-24 on the following page.
8. Insert the eyebolt in the hole on the other side of the chimney bracket as shown in Fig. 4-25 on the following page.
9. Place a nut on the eyebolt and tighten the nut until the bracket does not move.

63

■ **4-24** *Inserting the eyebolt on the bracket.* Thomson Consumer Electronics

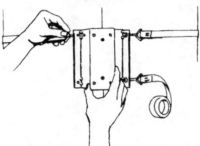

■ **4-25** *Mounting the bracket.* Thomson Consumer Electronics

10. Tighten both nuts.

11. Repeat steps 5 through 10 to attach the second metal strap.

Attaching the mounting foot to the DSS chimney mount

1. Align the mounting foot's four outside holes with the four holes on the chimney bracket.

2. Secure the mounting foot to the chimney bracket with the four bolts and nuts supplied in the DSS Chimney Mount kit.

3. Tighten all of the nuts and bolts. The finished installation should appear similar to the chimney mount in Fig. 4-26.

■ **4-26** *The finished chimney-mount installation.* Thomson Consumer Electronics

Vertically aligning the mast

After you attach the mounting foot to the surface you have selected, you need to vertically align the mast so that you can properly adjust the dish to receive the strongest signal. You can use a
plumb line or a bubble level.

If you selected a pole mount, you can skip this section because the
pole is the mast, and you aligned the pole during the installation.

1. Loosen the two bolts securing the mast to the mounting foot
   so the mast moves freely (refer to Fig. 4-27).

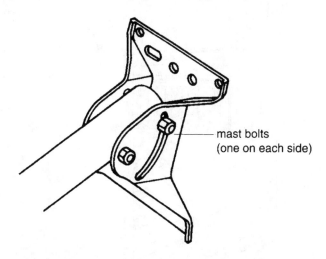

mast bolts
(one on each side)

■ **4-27** *The mounting foot adjustment.* Thomson Consumer
Electronics

2. Move the mast to the correct position for the location of the
   mounting foot, as shown in the example on the following page
   (Fig. 4-28).

3. Hold the bubble level or plumb line next to the mast, as shown
   in Fig. 4-29 on the following page. Gently move the mast so
   that the top portion aligns with the leveling tool.

4. When the top portion of the mast is vertically aligned, tighten
   the two bolts.

Securing the LNB support arm and dish to the mast

If you select a pole mount, the pole is the mast. If you attached the
mounting foot to a surface, mount the dish on the mounting foot's
mast.

1. If necessary, slightly loosen the three azimuth nuts on the
   support arm's mounting assembly (see Fig. 4-30 on the
   following page).

■ **4-28** *Various dish-mounting locations.*

Thomson Consumer Electronics

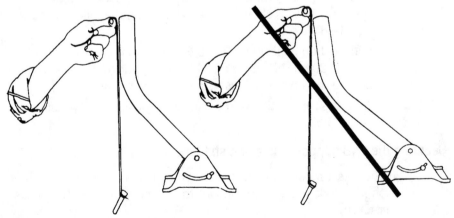

■ **4-29** *Correct and incorrect vertical alignment.* Thomson Consumer Electronics

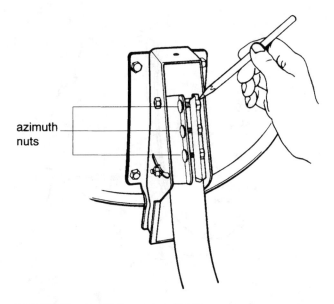

azimuth
nuts

■ **4-30** *The azimuth nut location.* Thomson Consumer Electronics

2. Slide the mounting assembly onto the mast until the top of the
   mast stops at the bolt at the top of the mounting assembly
   (see Fig. 4-31).

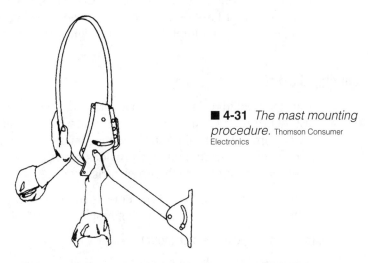

■ **4-31** *The mast mounting
procedure.* Thomson Consumer
Electronics

3. Using a plumb line or bubble level, rotate the LNB support
   arm assembly around the mast and confirm that top portion of
   the mast is vertically aligned in all directions.

4. Locate the scribe marks on the mast bracket on the LNB
   support arm on the following page (Fig. 4-32).

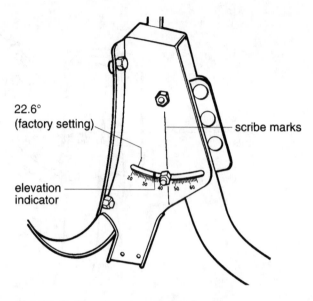

22.6°
(factory setting)

scribe marks

elevation
indicator

■ **4-32** *Setting the elevation indicator.* Thomson Consumer
Electronics

5. Confirm that the elevation indicator is at 22.6 degrees, the
factory setting.

6. Align the plumb line with the top of the scribe marks. If the
mounting foot has been properly installed and aligned, the
remainder of the plumb line should align with the scribe
marks.

Grounding the dish

Local electrical installation codes and the National Electrical Code
(NEC) both require the dish to be grounded. Grounding the dish
provides protection against static voltage buildup, which may damage equipment. Grounding also provides some protection against
surges induced by nearby lightning strikes.

In a pole mount, you ground the pole. In all of the other mounting
methods, you ground the mounting foot.

Grounding the pole in a pole mount

You will need the following materials:

☐ A ground rod (a copper-clad rod at least eight feet long is
best) and ground rod clamp

☐ ¼-20 × 1¾" zinc-plated bolt, ¼" star washer, and ¼" nut

☐ Ground wire

☐ Tie wraps

To ground the pole, follow these steps:

1. About one foot from the pole's cement base, drive the ground rod into the ground. According to the National Electrical Code, you must drive the ground rod eight feet into the ground.

2. Insert the ¼-20 × 1¾" zinc-plated bolt through the hole at the top of the pole (see Fig. 4-33).

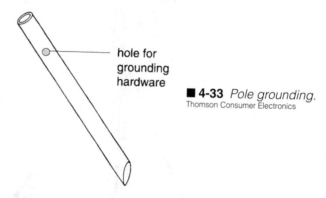

hole for
grounding
hardware

■ **4-33** *Pole grounding.*
Thomson Consumer Electronics

3. Insert the ¼" star washer then the ¼" nut onto the bolt.

4. Wrap the ground wire around the bolt next to the star washer and securely tighten the nut.

5. Route the ground wire down the pole and attach it to the ground rod using the ground rod clamp (see Fig. 4-34 on the following page).

6. Secure the ground wire to the pole using tie wraps.

Grounding the mounting foot

If you attached the mounting foot to a surface other than a pole, ground the dish by grounding the mounting foot.

You will need the following materials:

☐ Ground rod and ground rod clamp

☐ Ground wire

To ground the dish, follow these steps:

1. Drive the ground rod into the ground near the location where you attached the mounting foot. According to the National Electrical Code, you must drive the ground rod eight feet into the ground.

2. Insert the ¼-20 × ½" zinc-plated bolt supplied with the dish through the grounding hole on the mounting foot as shown in Fig. 4-35 on the following page.

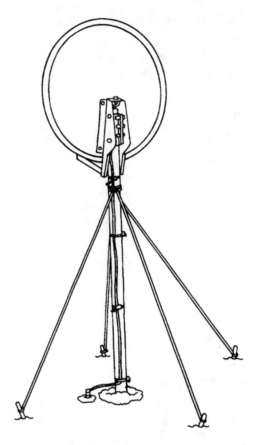

■ **4-34** *Grounding the dish to a ground rod.* Thomson Consumer Electronics

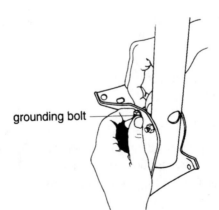

grounding bolt

■ **4-35** *The grounding bolt on the mounting foot.* Thomson Consumer Electronics

3. Insert the ¼" star washer, then the ¼" nut onto the bolt.

4. Wrap the ground wire around the bolt between the star washer and the surface of the mounting foot, and tighten the nut.

5. Route the ground wire down to the ground rod and attach it to the grounding rod using the ground rod clamp (see Fig. 4-36).

6. You can secure the ground wire to the mounting surface using insulated U-shaped tacks.

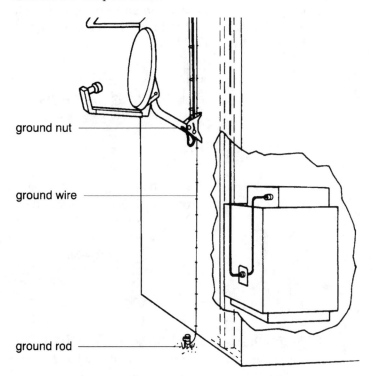

ground nut

ground wire

ground rod

■ **4-36** *DSS system grounding.* Thomson Consumer Electronics

Installing the grounding block

NOTE: Local electrical codes and the National Electrical Code (NEC) require a grounding block to be attached to the coaxial cable at a point of entry into the building to prevent damage from static voltage buildup and from surges induced by nearby lightning strikes.

Use the ground wire to connect the grounding block to a ground rod or a metal cold water pipe. Secure the ground wire to the ground rod using a ground rod clamp or to a metal cold water pipe using a cold water pipe clamp with the shortest possible length of wire. You need to check with your local electrical code for guidelines when using cold water pipes and ground rods for electrical grounding.

The grounding block has two coaxial cable connectors: one to connect the dish to the grounding block, and the other to connect the grounding block to the DSS receiver. To install the grounding block, perform the following steps:

1. Use the screws in the grounding block packet to attach the grounding block to the side of the house close to the cable's entry point (see Fig. 4-37).

■ **4-37** *Grounding block installation.* Thomson Consumer Electronics

2. Attach the coaxial cable to be routed to the dish to one connector on the grounding block, as shown in Fig. 4-38.

cable to dish

■ **4-38** *The coax from the dish connected to block.* Thomson Consumer Electronics

3. Attach the coaxial cable to be routed to the DSS receiver to the other connector on the grounding block as shown in Fig. 4-39.

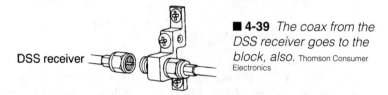

DSS receiver

■ **4-39** *The coax from the DSS receiver goes to the block, also.* Thomson Consumer Electronics

4. Insert the ground wire under the grounding screw and tighten the screw. See Fig. 4-40.

5. Attach the ground wire to the ground rod using the ground rod clamp (see Fig. 4-41).

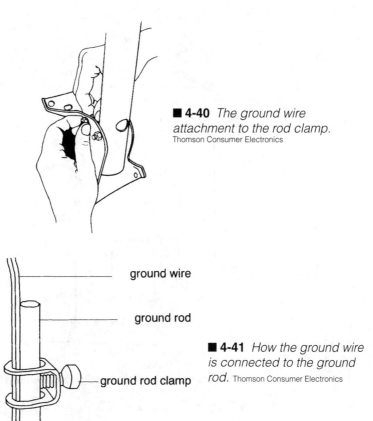

■ **4-40** *The ground wire attachment to the rod clamp.* Thomson Consumer Electronics

ground wire

ground rod

■ **4-41** *How the ground wire is connected to the ground* ground rod clamp *rod.* Thomson Consumer Electronics

Installing and routing cable

When you route the cable into your house, select the shortest route from the grounding block to the DSS receiver so that you do not use more than 112 feet of coaxial cable. If you need more cable, you can purchase more RG-6 coaxial cable with connectors attached. Also, you will need a TVRO bullet amplifier, which is attached to the end of the first cable before connecting any additional cable. The TVRO bullet amplifier boosts the satellite signal and overcomes the loss due to additional cable length.

CAUTION: When you install the cable and route it to the DSS receiver, be careful not to drill into hidden wiring, ducts, or plumbing pipes. DO NOT drill near an electrical outlet.

The example in Fig. 4-42 on the following page shows the cable being routed from the dish through the basement.

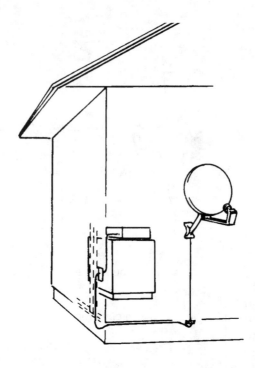

Cable routed from dish through a basement

■ **4-42** *Routing the cable through a base-ment.* Thomson Consumer Electronics

Routing cable through an exterior wall

If you are careful and know the location of electrical wiring, phone wiring, and the plumbing in your walls, you can route the coaxial cable through an exterior wall into your house.

☐ Make sure the hole you drill is above ground level.

☐ Determine the thickness and composition of the wall before you drill.

☐ Drill from the inside to the outside.

☐ If the exterior wall is near your DSS receiver, you can install a coaxial cable wall plate at the entry point. Then you can connect the receiver to the connector in the wall plate with coaxial cable (see Fig. 4-43).

☐ If the exterior wall is a basement or crawlspace wall, you can route the cable under the floor to the DSS receiver.

☐ Drill between studs just above the baseboard and at least 5 inches or more from any door or window opening. Otherwise, you could drill into framing.

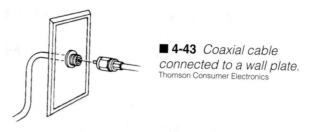

■ 4-43 *Coaxial cable connected to a wall plate.*
Thomson Consumer Electronics

When you are through with the installation, seal the entry points carefully with silicone sealant. You will need the following materials and tools:

☐ Two lengths of RG-6 coaxial cable. (If you plan to install a coaxial cable wall plate, you will need an additional length of coaxial cable.)

☐ Coaxial cable wall plate

☐ Tie wraps to secure the coaxial cable to a pole or mast

☐ Tape

☐ Silicone sealant

☐ An electric drill

☐ A screwdriver

☐ A ½" high-speed-steel bit 12 inches long, for drilling through interior walls or through an exterior wood-siding wall.

☐ A ½" masonry bit 12 inches long, for drilling through an exterior brick, poured cement, or cinder block wall.

To route cable through a wall, follow these steps.

1. Make sure the DSS receiver is unplugged.

2. Drill a ½" hole through the wall using a ½" high speed or masonry drill bit approximately 12 inches long.

3. Carefully cover the connector on the end of the cable with tape before pushing it through the wall. Make sure you do not bend the center wire.

4. Push the cable through the hole in the wall. Coaxial cable is stiff and usually can be pushed through a wall.

   NOTE: If the wall is a basement wall or crawl space and you need to route the cable under the floor to the DSS receiver, refer to "Routing cable under an interior floor" after step 9. If the wall is an exterior wall near the DSS receiver, continue with step 5.

5. Remove the tape and make sure the center wire on the connector is straight.

75

6. Attach the coaxial cable to the connector on the back of the coaxial cable wall plate.

7. Mount the wall plate on the wall using the screws supplied with the wall plate.

8. Connect the connector on the wall plate to the IN FROM SAT connector on the DSS receiver with coaxial cable.

9. Seal all entry points carefully using silicone sealant.

Routing cable under your interior floor

1. Make sure the DSS receiver is unplugged.

2. Follow previous steps 2 and 3. Then, route the cable across the basement or crawl space until you are below the room containing the DSS receiver.

3. Drill a ½" hole through the floor.

4. Push the cable up through the hole in the floor.

5. Pull any slack wire back into the room below and coil it neatly.

6. Carefully remove the tape from the connector and make sure the center wire on the connector is straight. Then, attach the connector to the IN FROM SAT connector on the DSS receiver.

7. If the room below is a basement or crawl space, attach the wire to the joists or exposed beams using cable clips.

Connecting the cable to the LNB

1. Locate the Phillips-head screw and special hex retainer nut in the hardware kit supplied in the DSS dish.

2. Thread coaxial cable from the grounding block through the mounting foot, mast, and LNB support arm. Leave a drip loop as shown in Fig. 4-44.

3. Attach the coaxial cable to the LNB and tighten the "F" connector as shown in Fig. 4-45.

Attaching the LNB to the LNB support arm

1. Slide the cable connection end of the LNB into the rectangular opening in the LNB support arm so that the LNB points up and toward the dish. Carefully align the LNB mounting holes.

2. Insert the special hex retainer nut into the LNB mounting hole on top of the LNB support arm.

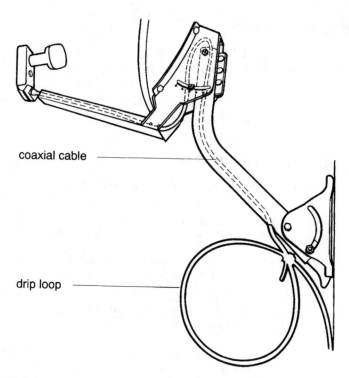

coaxial cable ——————————

drip loop ——————————

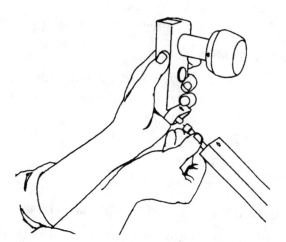

■ **4-45** *Attaching the coax cable to the LNB unit.* Thomson Consumer Electronics

3. Insert the Phillips-head screw into the LNB mounting hole and into the hex retainer nut on the bottom of the LNB support arm.

4. Securely tighten the screws.

Sealing the entry points

It is important to use silicone sealant to seal all cable entry points into the house and any test holes you drilled. The sealant prevents water from getting into your house. Later, if you need to replace the cable, you can remove the hardened silicone sealant, change the cable, and apply new sealant.

Pointing the satellite dish

CAUTION: AVOID POWER LINES! When installing the dish and connecting the satellite antenna connections, take extreme care to avoid contact with overhead power lines, lights, and power circuits. Contact with any or all of these things could be fatal.

The DSS dish must be pointed accurately toward the satellite to receive the satellite signal. When you point the dish, you set the elevation (move the dish in the up-and-down direction) then set the azimuth (move the dish left-to-right on the mast) until you receive the strongest signal.

Use the satellite location chart and a compass to help you locate the relative position of the satellite and point the dish. Then, to adjust the dish for the best signal reception, use the DSS receiver's on-screen signal strength meter. Information on using the receiver's on-screen display will be found in a later chapter.

Procedure for pointing the dish

There are four basic steps for pointing the dish:

1. Using the DSS receiver's on-screen menu system to find your elevation and azimuth settings.
2. Setting the elevation.
3. Setting the azimuth.
4. Fine-tuning the azimuth.

Using the menu system to find elevation and azimuth

Before you can use the menu system to find your elevation and azimuth, you must connect the DSS receiver to your TV receiver. When you have completed the installation, and are receiving the satellite signal, reconnect your DSS receiver, TV, and other components. During the dish installation, follow these steps to connect your TV set to the DSS receiver:

1. Connect the OUT TO TV connector on the DSS receiver to the TV's ANT IN connector with the coaxial cable supplied with the receiver (see connections in Fig. 4-46).

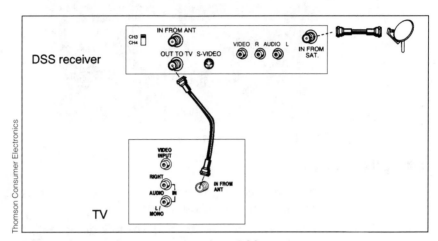

■ **4-46** *Connections on the rear of the DSS receiver.*

2. Turn ON the TV and the DSS receiver.
3. Tune the TV to channel 3 or 4, depending on the setting of the CH3/CH4 switch on the back of the receiver.
4. Press the DSS button on the remote control.
5. Press MENU on the remote control to bring up the main menu. (Each screen has instructions that will help you use the menu system.)
6. Select Options.
7. Select Setup.
8. Select Dish Pointing.
9. Select Point Dish Using Your Zip Code to bring up the Zip Code display (see Fig. 4-47 on the following page).
10. Enter your zip code using the number keys on the remote control. Select OK when you have finished. The display screen gives you the correct elevation and azimuth for your location. Write down both numbers so that you can refer to them later.

   AZIMUTH: _____   ELEVATION: _____
11. Select OK to bring up the Dish Pointing display screen.
12. Select Signal Meter to bring up the signal strength meter.

Leave the signal strength meter on the screen during the following steps.

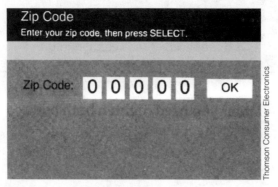

■ **4-47** *One of the menu screens.*

Setting the elevation

1. Loosen the nuts securing the two elevation bolts so that you can easily move the dish up and down.

2. Line up the elevation indicator with the tick mark corresponding to the elevation number given on the display screen. The elevation indicator, shown in Fig. 4-48, is the metal plate just behind the elevation tick marks.

3. When the indicator is aligned with the correct tick mark for the elevation you need, tighten both bolts.

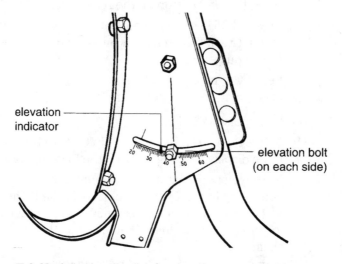

■ **4-48** *Adjusting the dish for elevation.*

Setting the azimuth

So far you have used the on-screen menu system to find the correct elevation and azimuth settings for your location, and you have set the elevation on the dish. Now you will set the azimuth. The following are the basic steps setting the azimuth:

1. Loosen the three azimuth nuts on the LNB support arm as shown in Fig. 4-49. You set the azimuth by moving the dish left-to-right, as shown in Fig. 4-50.

2. Refer to the map in Fig. 4-51 on the following page and point the dish in the general direction of the satellite, roughly south of Texas.

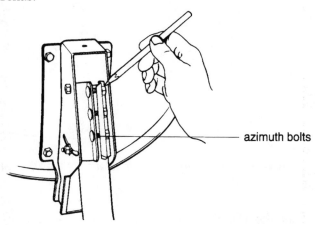

azimuth bolts

■ **4-49** *Setting the dish azimuth adjustment.* Thomson Consumer Electronics

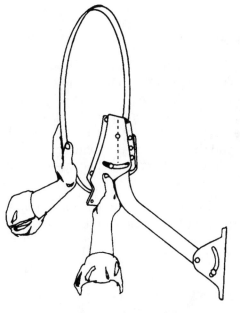

■ **4-50** *Set the azimuth by moving the dish from left to right.* Thomson Consumer Electronics

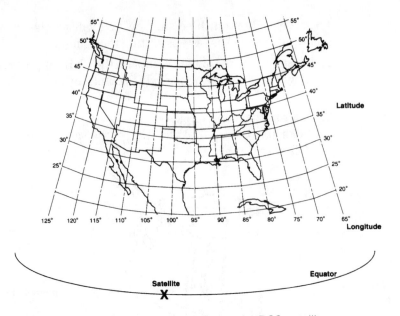

■ 4-51 *A map of the U.S. with relation to the DSS satellite.* Thomson Consumer Electronics

Fine-tuning the azimuth

Having set the elevation and pointed the dish in the general direction of the satellite, you will now use the on-screen signal strength meter to adjust the dish precisely and lock in on the satellite signal.

Using the signal strength meter

To help you lock in on the satellite signal, the on-screen signal strength meter provides an audible tone in addition to displaying the relative strength of the signal. As you point your dish toward the satellite, the signal strength meter produces short tones that indicate whether you have located the satellite. As you adjust the dish and the DSS receiver comes closer to locking in on the satellite signal, the short tones become one long, continuous tone.

Pointing the dish properly can take several minutes from the time you begin moving the dish left-to-right until you lock in on the satellite signal.

The following are some tips that can make this process easier for you:

☐ If you have wireless headphones, you can attach the headphones to the jack in the DSS receiver. Then you can hear the tones when you are away from the TV.

☐ If you do not have wireless headphones, you can turn up the TV's sound or have someone relay the information to you.

Locking in the satellite signal

1. At the top of the mast is a piece of tape with evenly spaced tick marks. Line up a scribe mark on the LNB support arm with one tick mark.

2. Gently turn the dish to the left or right, one tick mark at a time as shown in Fig. 4-52.

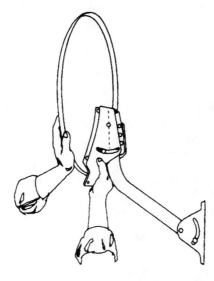

■ **4-52** *Fine-tune the dish direction one tick at a time after allowing the DSS receiver to complete two tuning cycles.* Thomson Consumer Electronics

3. Pause a few seconds while the receiver goes through its tuning cycles. The DSS receiver must go through two complete tuning cycles (two beeps) before the dish should be moved to the next tick mark. When the dish has locked in on the satellite signal, the short beep changes to one long, continuous tone.

4. Gently turn the dish one tick mark to the left. Pause and listen for the long, continuous tone.

5. Repeat step 4 until the continuous tone changes to short beeps. Then, mark the location of the scribe mark used in step 1 on the strip of tape on the mast.

6. Gently turn the dish one tick mark to the right. Pause and listen for the continuous tone.

7. Repeat step 6 until the continuous tone changes to short beeps. Then, mark the location on the scribe mark used in step 1 on the strip of tape on the mast. The strongest signal will be centered between the two places you marked.

8. Now, gently turn the dish so that the scribe mark used in step 1 is centered between the two places you marked in steps 5 and 7. The signal strength meter should indicate a strong signal. If needed, you can fine-tune the signal further by turning the dish to the left or right one tick mark and seeing if there is any change in the tone, or if the signal strength number gets larger.

   NOTE: The signal meter does not have to reach a reading of 100. Use the signal meter to find the highest possible signal

9. When you have locked in on the strongest possible signal, tighten the azimuth bolts securely, while being careful not to jar the dish out of position.

Fine-tuning the elevation

After you fine-tune the azimuth, you can use the same procedure to fine-tune the elevation. When you fine-tune the elevation, make sure you move the dish up or down one tick mark at a time, and listen for the tone after each movement.

If you have problems fine-tuning the dish, check the vertical alignment of the mast. If the mast is not perfectly vertical, you may not be able to fine-tune the dish properly.

Satellite location chart

You can use the longitude-latitude chart to set the elevation and azimuth more precisely than you can using the on-screen menu system.

1. Find your approximate location on the map of the United States shown in Fig. 4-53.

2. Locate the latitude (horizontal) line nearest your location.

3. Trace the latitude line to either side of the chart to find your approximate latitude and write it down. _____

4. Locate the longitude (vertical) line nearest your location.

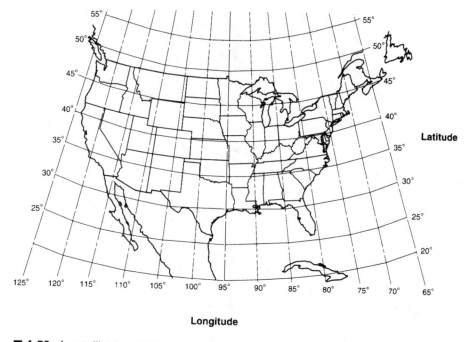

Longitude

■ **4-53** *A satellite location chart.* Thomson Consumer Electronics

5. Trace the longitude line to the bottom of the chart to find your approximate longitude and write it down. _____

6. Now refer to Fig. 4-54 on the following page and locate the latitude. Then, locate the longitude within the latitude that is nearest your geographic location.

7. Write down the azimuth number that corresponds to your latitude and longitude. _____

8. Write down the elevation number that corresponds to your latitude and longitude.

9. Hold your compass so that the dial lines up with north and south. Then, look in the direction that corresponds to the satellite location number. This is the direction in which your dish should point. The elevation number tells you where to set the elevation on the LNB support arm (refer to Fig. 4-55 on the following page).

Example

If you are located in St. Louis, Missouri, your latitude is almost 40 degrees. Your longitude is a little over 90 degrees. Looking at Fig.

Latitude	Longitude	Azimuth	Elevation	Latitude	Longitude	Azimuth	Elevation	Latitude	Longitude	Azimuth	Elevation
25	80	224	52.5	35	100	173	49.3	45	90	194	37.0
	85	214	55.7		105	154	49.1		95	183	37.8
	90	201	58.3		110	152	48.2		100	171	38.2
	95	188	60.0		115	143	46.7		105	161	36.9
	100	174	60.8		120	134	44.6		110	151	37.4
	105	163	60.4						115	144	36.3
	110	149	59.1	40	70	239	33.8		120	136	34.8
	115	137	56.8		75	228	36.6		125	130	33.0
					80	218	38.9				
30	80	221	48.2		85	207	40.9	50	65	246	23.4
	85	210	50.9		90	195	42.4		70	229	25.6
	90	198	53.0		95	183	43.3		75	228	27.6
	95	186	54.4		100	172	43.7		80	214	29.3
	100	174	56.2		105	162	43.6		85	206	30.7
	105	162	54.8		110	152	42.8		90	195	31.8
	110	151	53.7		115	143	41.5		95	184	32.4
	115	141	51.8		120	135	39.8		100	172	32.7
									105	161	32.6
35	75	227	40.9	45	65	246	27.2		110	151	32.1
	80	219	43.6		70	238	29.8		115	142	31.2
	85	208	45.9		75	228	32.1		120	135	29.9
	90	196	47.7		80	217	34.1		125	128	29.3
	95	184	48.9		85	206	35.8				

■ **4-54** *A satellite location table.* Thomson Consumer Electronics

Compass

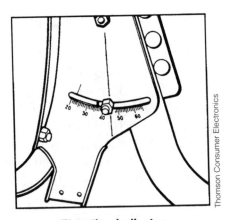

Elevation Indicator

Thomson Consumer Electronics

■ **4-55** *Using the compass and elevation indicator for dish direction setting.*

4-54, you see that a latitude of 40 degrees and a longitude of 90 degrees corresponds to an azimuth number of 195 and an elevation of 42.4. The technician shown in Fig. 4-56 is setting the elevation and aiming the DSS dish.

■ **4-56** *A technician setting the elevation of a DSS dish.*

DSS installation glossary

Alignment The process of adjusting the satellite dish to receive the strongest satellite signal.

Azimuth Left and right adjustments to the satellite dish. Technically, the degrees of rotation on a compass, measured in a clockwise direction from true North. This information can be used to determine the location of the satellite relative to your home and help you to point the dish toward the satellite.

Bullet amplifier A small device used to increase signal power (gain) and offset any signal loss caused by the coaxial cable and signal splitting devices. See TVRO amplifier.

Coaxial cable A type of cable used to transmit high-frequency signals. High-bandwidth cable that carries the satellite signal from the LNB to the DSS receiver. The DSS system uses RG-6 cable to connect the satellite dish and the DSS receiver. **NOTE:** There are many grades and brands of RG-6 cable.

Dish The part of the satellite dish antenna that collects, reflects, and focuses the satellite signal into the LNB.

Drip loop Several inches of slack in a cable that prevents water from collecting on the cable or running along the surface of the cable. A drip loop between the LNB and the entry point into a building also allows free movement of the dish while it is being adjusted.

DSS Digital Satellite System.

DSS receiver Receives, processes, and converts the satellite signal into picture and sound.

Earth ground Conducting connection to the earth for an electrical charge so that the electrical charge is at zero potential with respect to the earth.

Elevation Up-and-down adjustments of the DSS dish. Technically, the vertical angle that is measured from the horizon up to the satellite. This information helps you locate the satellite and point the dish toward it.

"F" connector A special type of connector used commonly to terminate the RG-6 coaxial cable.

Feedhorn The input of the LNB. This collects and focuses satellite signals reflected by the dish.

Grounding block A device that connects two coaxial cables and which can be grounded to earth to prevent electrical surges through the coaxial cables.

Ground rod Metal pole, typically eight feet long, driven into the ground to connect an electrical current to earth.

Ground wire A wire connecting an electrical circuit to a ground rod.

LNB Low Noise Block converter. This unit is mounted at the focal point of the dish, and is used to amplify and convert satellite signals into frequencies sent to the DSS tuner.

Latitude The distance, measured in degrees, between a location on the surface of the earth and the equator.

Longitude The distance, measured in degrees, between a position on the earth and the prime meridian.

Main menu The first menu in the on-screen menu system. Menus are lists of choices that allow you to customize the DSS receiver and access features available through the on-screen menu system.

Mast A metal pipe attached to the mounting foot and supporting the LNB support arm and dish. In a pole mount, the metal pole is the mast.

Mounting foot A DSS dish assembly that attaches to the mounting surface and mast.

88

Plumb line A string with an attached weight that provides a reference line perpendicular to the ground.

Preterminated coaxial cable Cable with factory-installed weatherproof "F" connectors.

Rafters Timbers, usually spaced 16 inches apart, used to support a roof.

Signal meter An on-screen meter accessed through the main menu, which displays the relative strength of the satellite signal that is used to fine-tune the dish. The signal meter also provides auditory feedback.

Stud Vertical timbers, 2×4 or 2×6, usually spaced 16 inches apart and used to support a wall.

TVRO amplifier TeleVision Receiver Only bullet amplifier used to overcome signal loss in a long coax cable run.

Information in this chapter is courtesy of Thomson Consumer Electronics.

Complex "SMATV" distribution system

5

THIS CHAPTER WILL COVER THE DSS NEW HOME PREWIRING and distribution system. It will contain information on "SMATV" (Satellite Master Antenna TeleVision). Off-air antenna placement, satellite dish placement, new home prewiring, safety information, and various hookup configurations are also addressed in this chapter.

The SMATV system

Basically, this type of system is a privately owned and operated antenna system that is designed to receive and distribute satellite, "off-air" (broadcast) TV signals, and/or standard cable signals. These signals are received, combined, and then distributed to multiple TV receivers within a structure via a coaxial cabling network. An SMATV system can be as simple as those found in a private residence or as complicated as those found in motels, apartment buildings, or schools, etc. In order to distribute the signals without a noticeable loss of signal quality, an SMATV system must be carefully planned and installed through the effective use of the proper equipment and installation techniques. Another commonly used and probably better-known acronym is MATV. An MATV system (Master Antenna TeleVision) is basically very similar to an SMATV system, except that it is primarily designed to receive and distribute off-air or standard broadcast signals only.

An SMATV system can be divided into two major and distinct sections—the head end and the distribution system. The head end consists of three basic elements:

1. Antenna installation
2. Signal processing equipment
3. Amplifiers (if needed)

The second major section of an SMATV system is the cabling or distribution system. A well-designed and installed distribution system is necessary in order to provide acceptable signal levels to every receiver in the system. The distribution system should provide a "clean" signal to the sets by isolating each receiver from the system, and should also deliver the proper amount of signal to each receiver. This portion of the system consists of trunk lines, splitters, feeder lines, and signal splitters. Some of the other equipment in the distribution system includes (but is not limited to) such devices as line taps, TV signal wall outlets, combiners, etc. (refer to the diagram of the distribution system in Fig. 5-1).

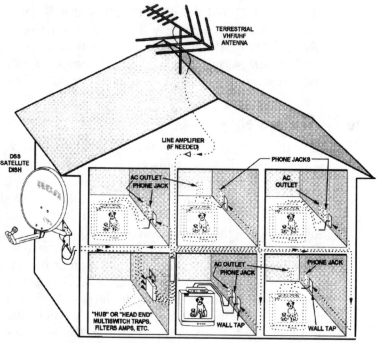

■ **5-1** *The SMATV distribution system.*

It is very important to carefully design and layout the system prior to beginning the installation. The requirements of the system will determine the type of equipment that is needed for the head end. The actual design and the difficulty of the installation depends largely on the dimensions and signal requirements of the building in which the system is installed.

NOTE: For the self-installer: The design and installation of a complex SMATV system can be somewhat intimidating and difficult for the nontechnical. Depending on the system and its complexity, you may save time and money by consulting a professional to design and install your SMATV system.

Off-air antenna placement

The off-air terrestrial antenna is a critical component of any SMATV system, and is responsible for receiving the off-air broadcast signal. For this reason, the ultimate quality of the broadcast TV reception can be no better than the quality of the signal from the off-air antenna. It is vital that the off-air antenna be selected and installed with care. Some installations will use only one antenna that can receive all channels (both VHF and UHF) and is referred to as a *broadband antenna*. However, if the TV broadcast stations or channels to be received lie in different directions, or if adjacent channel rejection is a problem, a *single-channel antenna* may be selected. The number of channels to be received, the direction and distance to the transmitters, the type of signal (UHF or VHF), and the available signal level all must be considered when choosing an off-air antenna. There are numerous publications that can help with the selection of the proper off-air antenna selection.

Satellite dish placement

The selection of a mounting location for the RCA brand DSS satellite dish in an SMATV system installation requires special attention to avoid any potential obstructions. This can include such things as trees, hills, or other buildings that are in the "Line-of-Sight" to the satellite. It is also very important to mount the dish correctly in order to minimize wind loading and aiming problems. To select the best mounting location and install the DSS dish properly, it is very important to carefully follow the instructions that are found in chapter 4 of this book.

New home prewiring

During the construction of a new home, it is more efficient and cost-effective to prewire the house for all of the off-air, satellite, and cable equipment that might be installed once construction is completed. All prewired cabling should be finished prior to the installation of the interior drywall or paneling. This way all cabling and outlet boxes can be routed or installed in much the same manner as the electrical wiring. Consideration should also be given to prewiring each room for telephone hookup, in the event that each room has its own DSS receiver. The ideal situation would be that the TV signal outlet, the telephone jack, and an ac outlet be in close proximity to the TV and DSS receiver (see Fig. 5-2).

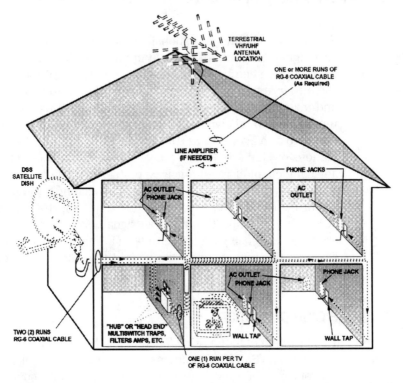

■ 5-2 *Prewiring a new home during construction (Home Run method).*

However, to effectively prewire a home during construction, there are several important factors that must be considered. These decisions should be made prior to starting the installation. These six factors are:

1. The type of system to be installed. This could be a single or multiple TV/DSS receiver system, etc.

2. The placement of the satellite dish on the exterior of structure.

3. The selection and location of the off-air (terrestrial) antenna.

4. The location of the hub or head-end equipment (such as amps, multiswitches, filters, etc.)

5. How cabling is to be routed through the structure (in walls, the attic, the crawl space, etc.)

6. The location of TV signal wall plates, phone jacks, and ac outlets.

NOTE: If the self-installer or homeowner is unfamiliar and inexperienced in the design and installation of RF distribution systems, it is advisable to obtain the assistance and/or advice from a professional installer prior to starting the installation.

The Home Run system

The most popular type of prewire residential RF distribution system utilized today is the Home Run system shown in (Figs. 5-2 and 5-3). The Home Run system is a parallel output system. In this type of system, each cable outlet has its own dedicated output from the hub or head-end equipment. This is no doubt the best system in terms of quality signal distribution. The Home Run system also provides the greatest flexibility for future expansion. Adding to the system is as simple as connecting an additional DSS receiver and plugging into the phone line. There are other types of distribution systems; however, they are not recommended because of their limitations.

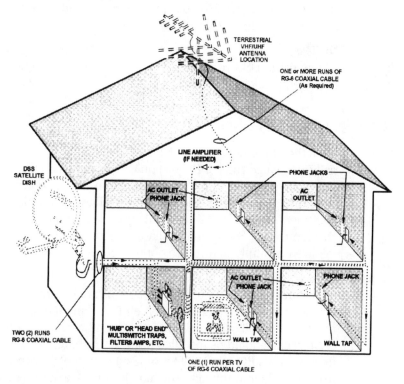

TERRESTRIAL VHF/UHF ANTENNA LOCATION

ONE or MORE RUNS OF RG-6 COAXIAL CABLE (As Required)

LINE AMPLIFIER (IF NEEDED)

PHONE JACKS

DSS SATELLITE DISH

AC OUTLET PHONE JACK

AC OUTLET

AC OUTLET PHONE JACK

PHONE JACK

TWO (2) RUNS RG-6 COAXIAL CABLE

"HUB" OR "HEAD END" MULTISWITCH TRAPS, FILTERS AMPS, ETC.

WALL TAP

WALL TAP

ONE (1) RUN PER TV OF RG-6 COAXIAL CABLE

■ **5-3** *New home construction coax wiring.*

Installation and design tips

When doing the basic design work and layout of the system, consideration should be given to other types of wiring, such as an alarm and security system, motion detector system wiring, or data communications lines.

Location of the satellite dish on the structure is very important in that you must have an unobstructed line-of-sight view to the satellite at which you are aiming. Another major reason for predetermining the location of the dish is the routing of the cabling from the satellite dish to the hub or head end. This allows the coax runs for the satellite dish to be installed prior to the walls being finished. For optimum performance, the cable run from the dish to the head end (or DSS receiver) should not exceed 100 feet. If the cable run is greater than 100 feet, a special amplifier may be needed in order to boost the signal level from the DSS dish. It is important to place all amplifiers in an area or location where they will not be walled over during construction. This is so they are accessible for service at a later time. Again, RG-6 is the recommended coax between the DSS dish and the hub or DSS receiver.

NOTE: Never use RG-59 coax for the cable run between the DSS dish and the DSS receiver, because of that cable's characteristic high-frequency loss.

Coaxial cable tips

Even though only a "Single LNB" satellite dish may be installed initially, two cable runs of RG-6 should be installed during prewiring. This allows for easy upgrading and expansion of the system at a later time. Because of the special installation and alignment requirements of the DSS satellite dish, it is very important that all instructions in DSS system installation section found in chapter 4 be followed carefully in order to obtain optimum signal reception.

Predetermining the location of the off-air (terrestrial) antenna is important in order to obtain an adequate broadcast signal. This enables you to determine the type of antenna mount that will be needed. Also, all other cabling from the off-air antenna to the hub or head end should be installed prior to finishing the interior walls. In most cases RG-59 (100 feet or less) is suitable for the cable run from the off-air antenna to the hub. In some cases where the cable run is extremely long or there is excessive high-frequency signal loss (UHF bands), it may be necessary to install RG-6 coax cable and/or line amplifiers. It is important to read and follow all mounting and installation instructions that are usually packed with off-air antennas.

Hub/head systems

The hub or head end of the system is the central location where all the cabling comes together (note Fig. 5-3). The location of the hub is very important and should be easily accessible after all construction is finished. This is for ease of final installation and also in the event that the equipment needs to be serviced in the future. It should be easily accessible in the event that the equipment should require upgrading or expansion. Some typical locations for the hub could be an attached garage, an unused closet, or possibly a utility room or cabinet. Areas such as attics should be avoided, as accessibility may be a problem. Crawl spaces may not be good because of moisture, which could lead to corrosion of the cable connectors and other equipment.

Let's now look at how to route the cable through the building. There are, of course, many ways to install the cabling. A couple of options are to route the cabling through the attic or crawl space. These are not the most desirable locations, for the same reason that the head end equipment should not be installed in these types of areas (inaccessibility and corrosion due to moisture).

The most desirable method is to install the cabling in the same way that the electrical wiring is installed, in the walls prior to hanging the drywall. When installing the cabling, DO NOT USE A STAPLE GUN to attach the cable to the 2 × 4 studs or other parts of the structure. Instead, use cable clips; they will not puncture or damage the cabling. Also, when installing extra long cable runs (greater than 100 feet), a line amplifier may be needed at the beginning of the run, and its location should be selected with care so it is accessible at a later time.

The last point to cover is the location of the outlets for the TV signal. These should be chosen carefully, and should be relatively close to where the receiver will be placed. As mentioned earlier, the TV outlet should be installed close to a telephone jack and an ac outlet. For example, installing the TV outlet and ac outlet on one side of the room and the telephone jack on the other side is not appropriate for a DSS system installation. However, in the event that the telephone jack is located a considerable distance from the TV and ac outlets, or there is no telephone jack present, the RCA wireless phone jack (MODEL #D916) may be used to provide communications for the DSS system.

NOTE: As mentioned earlier, the design and installation of an RF distribution system can prove difficult for those not experienced in this field. For this reason, if questions or problems are encountered, consult a professional.

Safety information

DANGER—AVOID POWERLINES. The outdoor dish antenna used to receive satellite signals and the cable used to connect the outdoor dish antenna to the indoor receiving unit are required to comply with local installation codes and the appropriate sections of the National Electrical Code (NEC), especially Article 820. These codes require proper grounding of the metal structure of the outdoor dish antenna and grounding of the connecting cable at a point where it enters the house (or other building).

If you are having a professional installer make the installation, the installer must observe installation codes while making the dish installation. The RCA DSS do-it-yourself installation kit contains instructions on how to make the installation in compliance with the National Electrical Code (NEC). If additional local installation codes apply, contact the local inspection authorities.

National Electrical Code compliance and other restrictions

Before installing the DSS system, check the electrical code guidelines in your area. Also check the zoning codes, covenants, and community restrictions in your area. Some rules prohibit installing large satellite dishes, but may allow small ones (the RCA DSS System utilizes an 18-inch satellite dish). Also, there may be restrictions in your area that limit the mounting height of dishes.

If you encounter homeowner or community restrictions, call 1-800-679-4776. Personnel at this number can provide information that may be helpful when attempting to obtain permission to install a DSS System on your property.

DSS hookup configurations

This section is intended to familiarize you with some different hookup configurations. These examples do not represent all the different possible combinations, but are intended only as a guide in planning your system. For example, these illustrations do not include such things as VCR or laserdisc hookups. Also, cable hookup is not shown; however, a cable TV input could be substituted for the off-air antenna.

Single TV/single receiver/single output LNB

A drawing of this equipment is shown in Fig. 5-4. You will need the following equipment and materials:

1. A DSS system with a single receiver (single output LNB).
2. A VHF/UHF antenna.
3. A VHF/UHF amplifier.
4. Miscellaneous cabling (RF/audio/video/S-VHS).
5. A satellite amplifier—(RCA #D903 bullet type).

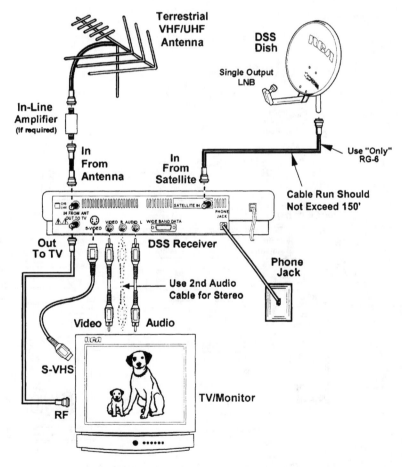

■ 5-4 *Single-TV/single-receiver/single-output LNB.*

Benefits and/or restrictions

1. Simple/cost-effective installation.
2. Only a single phone jack is required.
3. It does not require hub installation, but is difficult to expand.
4. Single TV hookup only.

Dual TV/dual receiver/dual output LNB

An illustration of this equipment setup is shown in Fig. 5-5. The following equipment and materials will be required:

1. A DSS advanced system and second receiver (the advanced system includes a dual-output LNB).
2. A VHF/UHF antenna.
3. A line splitter (RCA #D918).
4. Miscellaneous cabling (RF/audio/video/S-VHS).

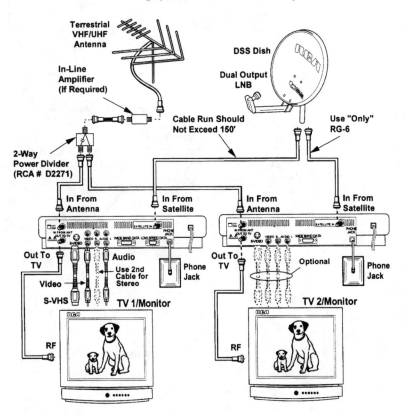

■ **5-5** *Dual-TV/dual-receiver/dual-output LNB.*

5. A VHF/UHF line amplifier.

6. A satellite amplifier, if it is required (RCA #D903 bullet).

Benefits and/or restrictions

1. Simple/cost-effective installation.

2. It does not require hub installation, but is difficult to expand.

3. It requires a phone jack at each receiver location.

Multiple TV/multiple receiver/dual output LNB with multiswitch

The complete drawing of this equipment setup is shown in Fig. 5-6. The following equipment and materials will be required:

1. A DSS system (dual-output LNB) with additional receivers (model #DRD102RW).

2. DIPlexers.

3. A VHF/UHF antenna.

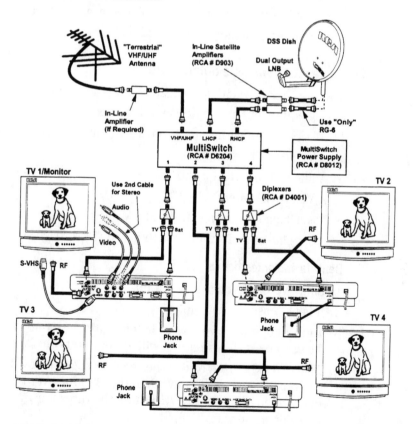

■ **5-6** *Multiple-TV/multiple-receiver/dual-output LNB with multiswitch.*

4. A VHF/UHF line amplifier.

5. Miscellaneous cabling (RF/Audio/Video/S-VHS, if required).

6. A multiswitch (nonamplified/amplified terrestrial).

7. A satellite amplifier (RCA D903) if it is needed.

Benefits and/or restrictions

1. The system is flexible, versatile, and expandable. It is an efficient system for using prewired new-home construction.

2. Each TV can view VHF/UHF or satellite signals independently of each other.

3. It requires extensive installation skills and knowledge of RF distribution systems.

4. For four or more outputs, use a multiple multiswitch installation (refer to Fig. 5-9).

5. It requires installation of a hub in a central location.

6. It requires a phone jack at each receiver location.

Dual TV/single receiver/single output LNB

A drawing of this equipment setup is shown in Fig. 5-7. The following equipment and materials will be necessary:

1. A DSS system with a single receiver (single output LNB).

2. A VHF/UHF line amplifier.

3. A line splitter (RCA # D918).

4. A VHF/UHF antenna.

5. Miscellaneous cabling (RF Audio/Video/S-VHS).

6. A satellite amplifier, if one is required (RCA #D903).

Benefits and/or restriction

1. Simple/cost effective installation.

2. It is not easily expandable.

3. You can operate two TVs independently with one receiver. The same satellite channel will appear on both TV receivers.

4. The TV/DSS button routes the receiver signal to TV 2. TV 2 can receive VHF/UHF only when the DSS receiver is off. TV 1 may switch between VHF/UHF (via the tuner) or the DSS receiver signal via A/V input.

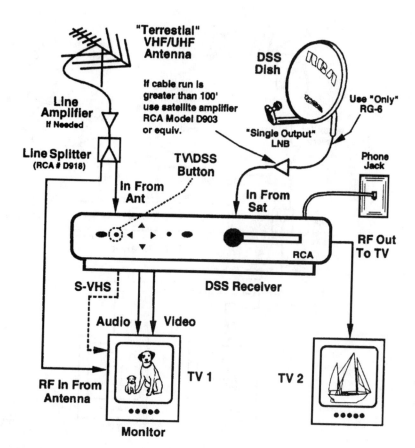

5-7 *Dual-TV/single-receiver/single-output LNB.*

Multiple TV/single receiver/single output LNB/MATV back feed

The layout for this system is shown in Fig. 5-8 on the following page. The following equipment and materials will be required:

1. A DSS system with a single receiver (single-output LNB).
2. An RF modulator (unused terrestrial channel).
3. A line splitter (RCA #D904).
4. An RF combiner (RCA #D918).
5. A VHF/UHF antenna.
6. Miscellaneous cabling (RF/audio/video/S-VHS).
7. A VHF/UHF line amplifier.
8. A satellite amplifier, if one is required (RCA #D903).

103

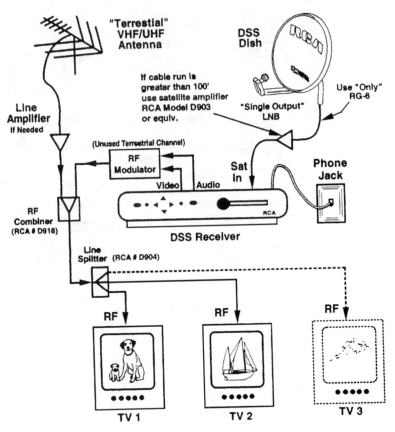

■ **5-8** *Multiple-TV/single-receiver/single-output LNB/MATV backfeed.*

Benefits and/or restrictions

1. It is difficult to expand.

2. Only a single phone jack is required.

3. Crosstalk interference is possible.

4. This system is least desirable, due to limitations and possible interference problems.

5. The output of the VHF/UHF antenna and the RF modulator must match closely, or the antenna could radiate RF energy.

6. It requires extensive installation skills and knowledge of RF distribution systems.

Multiple multiswitch installation

The drawing in Fig. 5-9 shows how to wire up the multiple-multiswitch hookup of a DSS dish terrestrial antenna, DSS receiver, and TV sets.

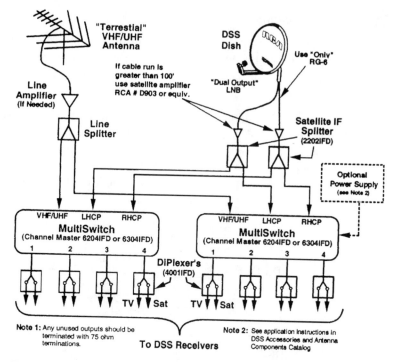

■ **5-9** *Multiple multiswitch installation.*

Off-air antenna reference information

This section includes some information that may help you with off-air antenna selections. Only a TV antenna professional can select and precisely install the correct antenna and assure you years of good TV/FM reception.

Back in the late 1940s, TV antennas were actually manufactured by hand at the shop or installation site. In fact, this was your author's job when first starting out in electronics. Each antenna was assembled, piece by piece, at the customers home. Also, the TV masts were made up of 1-inch and 1½-inch water pipe slipped inside each other, and then bolted together. It was quite a trick in the winter time on a steep, slick roof in Iowa. Now TV antenna manufacturers have eliminated this time-consuming labor by making preassembled TV antennas that provides excellent reception in most areas at reasonable costs.

Today, medium- to high-gain, broadband antennas are essential for good reception. Even the most expensive color TV or FM stereo receiver cannot perform to full potential without the strong signals that only a quality antenna system can provide.

Antenna specifications

Sensitivity ratings are based on optimum conditions over unobstructed terrain. What lies between a transmitter and an antenna installation will have a direct bearing on what type of antenna is appropriate.

Factors to consider are the power output and height of the transmitting antenna towers of the stations you wish to receive, the type of terrain between the towers and the receiving antenna, and the size and number of buildings that lie in the paths of the transmissions. The principal parts of a basic TV antenna are shown in Fig. 5-10.

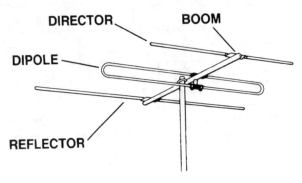

■ **5-10** *The principal parts of a basic TV antenna.*

Generally, VHF/FM and UHF/VHF/FM antennas have sensitivity classifications such as "fringe," "suburban," "deep fringe," etc. These classifications are designed to indicate at what distance from a TV transmitter the antenna will provide the best reception. Let's look at these antenna classifications.

Designation	VHF	UHF
Deepest fringe	100+ miles	60+ miles
Deep fringe	100 miles	60 miles
Fringe	80 miles	45 miles
Near fringe	60 miles	35 miles
Far suburban	50 miles	30 miles
Suburban	45 miles	35 miles
Far metropolitan	30 miles	25 miles
Metropolitan	25 miles	15 miles

Off-the-air antenna recommendations

These recommendations are listed in alphabetical order, and include market-specific reception comments in most cases. Recommendations are Channel Master brand antennas, unless specified otherwise.

Each market is defined by one of four letters: G, P, or X. Refer to the market description matrix for specifics. Several noteworthy items within the recommendations matrix:

☐ "WS," "WG," and "SF" model prefixes represent Winegard brand antennas.

☐ A "*" signifies that only the mandatory five channels may be received with the specified antenna solution.

☐ A "+" signifies a market where more than one antenna is required to receive the mandatory five channels.

☐ An "O" model suffix signifies that the recommended antenna is capable of outdoor mounting only, primarily due to size constraints.

Market description matrix

The market type description for off-the-air TV antenna recommendations will be found on the following page in Table 5-1. Table 5-2 on the following page gives the off-the-air antenna recommendations for most large U.S. cities. Table 5-3 on the following page gives you TV frequency and channel information.

Information, drawings and tables in this chapter is courtesy of Thomson Consumer Electronics. DSS is an official trademark of DirecTV, Inc., a unit of GM Hughes Electronics.

■ Table 5-1 Symbols found in off-air recommendation charts.

Market description matrix

Market type/Symbol		Description	Hardware	Installation*
Generic	G	- Clustered network transmitters; all five close together	- Channel Master #3016 - Misc. hardware, including: all mounting brackets, etc. - all cable and connectors	- Basic installation, includes: - Hookup to 2 TVs - Indoor attic-mounted - Average: 1.5 hours
Generic-plus	P	- Generally clustered network transmitters - Typically one transmitter in a remote location	- C/M #3016 + second bay antenna for outlying channel - Misc. hardware similar to "G" market	- More complex than "G" market - 1.8–2.4 hours range - Average: 2.2 hours
Area special	S	- Multi-positioned farm - Alternate solution is using a rotor with a broad band antenna - High volume specialty markets keep these special antenna prices competitive vs. a rotor solution	- Special antenna solutions identified and marketed - Misc. hardware similar to "G" market	- 1.5–3.1 hours range - Average: 2.4 hours
Open market	X	- Remote, multi-markets - Multi-positioned farms - Few channels typically available - Most solutions are amplified - Expensive, off-air solutions	- No reasonable indoor attic-mounted off-air solution - Alternate solutions need to be explored on per-market basis	

* Off-air antenna installations are indoor, attic-mounted and performed at the time of the DSS site installation.

■ Table 5-2   Antennas recommended for many U.S. cities.

Off-air antenna recommendations

Market name	Antenna Mkt type	Vendor/model	By market reception comments and restrictions
Alabama			
Birmingham, AL	G	3016	
Birmingham, AL	S	4700	
Decatur/Huntsville, AL	G	4225	
Huntsville, AL	P	4193	
Huntsville, AL	P	4225+	5 chs available; all UHF
Mobile, AL	G	3016	Rec. 7 chs
Arizona			
Phoenix, AZ	G	3016	9 chs available
Tucson, AZ	P	3016+	May need jointenna for 3016; ch 11
Tucson, AZ	P	3603	VHF yagi for ch 11; may need amp
Arkansas			
Little Rock, AR	S	WS-1801 O	Outdoor mount only
California			
Fresno, CA	G	4221	5 chs available
Los Angeles, CA	G	3016	Rec. 9 chs, incl KTLA
Sacramento/Stockton, CA	G	3016	
San Diego, CA	X		
San Francisco, CA	G	3016	Rec. 6 chs
Colorado			
Colorado Springs/Pueblo, CO	G	3016	5 chs available; issue w/security
Denver, CO	G	3016	Rec. 8 chs, incl KWGN
Connecticut			
Hartford/New Haven, CT	P	3016+	5 chs available; may require
Hartford/New Haven, CT	P	4308+	Ch 30, 61 UHF traps on 3016
Hartford/New Haven, CT	P	1473	VHF yagi for ch 8
Springfield, MA/Holyoke/Hartford, CT	X		
District of Columbia			
Washington, D.C.	G	3016	D.C. chs only; Baltimore chs w/rotor
Florida			
Fort Myers/Naples, FL	P	3016+	5 chs available
Fort Myers/Naples, FL	P	4308	UHF yagi for ch 26
Jacksonville, FL	G	3016	
Miami, FL	S	4693	
Miami, FL	S	WS-1648	
Miami, FL	S	WS-1771	
Miami, FL	S	WS-1774	
Orlando/Melbourne, FL	P	3016+	Rec. 5 chs; excludes Daytona
Orlando/Melbourne, FL	P	3603	VHF yagi for ch 2; no jointenna required
Tampa, FL	S	4701	
Tampa, FL	S	4710	
Tampa, FL	S	4792	
West Palm Beach/Fort Pierce/ Vero Beach, FL	P	3016+	5 chs available
West Palm Beach/Fort Pierce/ Vero Beach, FL	P	4221	Potential range issue at Ft. Lauderdale

Market name	Antenna Mkt type	Vendor/model	By market reception comments and restrictions
Georgia			
Atlanta, GA	G	3016	Rec. 6 chs, incl WTBS; not WATL
Savannah, GA	P	3016+	5 chs available
Savannah, GA	P	1473	VHF yagi for ch 9; may need jointenna
Illinois			
Chicago, IL	G	3016	Rec. 7 chs. incl WGN
Harrisburg, IL/Paducah KY/ Cape Girardeau, MO	X		
Peoria, IL	P	4221+	5 chs available; all UHF
Peoria, IL	P	4194	UHF yagi cor ch 43
Rock Island/Moline, IL/Davenport, IA	S	X	
Springfield, IL	P	4221+	4 UHF chs available +
Springfield, IL	P	3603	VHF yagi for ch 12; no jointenna needed
Indiana			
Evansville, IN	P	3604+	VHF yagi; all local VHF chs
Evansville, IN	P	4221	UHF yagi; all local UHF chs
Fort Wayne, IN	G	4221	Rec. local chs; no Indpls chs
Indianapolis, IN	S	WS-1573	
Indianapolis, IN	S	WS-1574 O	Outdoor mount only
South Bend, IN	G	4221	Local chs only no Indy/Chicago chs
Iowa			
Cedar Rapids/Waterloo, IA	S	X	
Davenport, IA/Rock Island/ Moline, IL	S	X	
Des Moines, IA	G	3016	5 chs available
Kansas			
Hutchinson/Wichita, KS	P	3016+	Some issue with Hutchinson area
Hutchinson/Wichita, KS	P	1473	May need ch 8, 12 yagi traps for
Wichita, KS	G	3016	3016
Kentucky			
Lexington, KY	P	4225+	5 chs available; all UHF
Lexington, KY	P	4193	
Louisville, KY	P	3016+	
Louisville, KY	P	3603	VHF yagi for ch 3
Louisville, KY	S	X	
Paducah, KY/Cape Girardeau, MO/ Harrisburg, IL	X		
Louisiana			
Baton Rouge, LA	G	3016	5 chs available
New Orleans, LA	P	3603+	UHF yagi; all local VHF chs
New Orleans, LA	P	4308	UHF yagi; all local UHF chs
Shreveport, LA/Texarkana, TX	G	3016	5 chs available
Maine			
Portland/Augusta/Lewistown, ME	G	3016	Some issue with Berlin area
Portland/Augusta/Lewistown, ME	P	3016+	Addresses midpoint of
Portland/Augusta/Lewistown, ME	P	1473	these 3 cities, incl Berlin

Market name	Antenna Mkt type	Vendor/model	By market reception comments and restrictions
Maryland			
Baltimore, MD	P	3016+	Rec. 5 Baltimore chs
Baltimore, MD	P	4193	D.C. chs w/rotor
Massachusetts			
Boston, MA	G	3016	Rec. 7 chs, incl WSBK
Michigan			
Detroit, MI	G	3016	Rec. 6 chs, not CBET (9)
Flint/Saginaw/Bay City, MI	S		WS-1785
Grand Rapids/Kalamazoo/ Battle Creek, MI	S	4673	
Grand Rapids/Kalamazoo/ Battle Creek, MI	S	WS-1705 O	Outdoor mount only
Lansing/Onodaga, MI	X		
Minnesota			
Minneapolis, MN	G	3016	Rec. 8 chs; some issue between metros
Mississippi			
Jackson/Vicksburg, MS	G	3016	2 Metros ok; some issue in-between
Missouri			
Cape Girardeau, MO/Paducah, KY/ Harrisburg, IL	X		
Kansas City, MO	G	3016	Rec. Kansas City chs only
Springfield, MO	S	WS-1734	
Springfield, MO	S	WS-1773 O	Outdoor mount only
St. Louis, MO	G	3016	6 chs available
Nebraska			
Lincoln/Hastings/Kearney, NE	P	3603+	5 chs available without AM
Lincoln/Hastings/Kearney, NE	P	3603	Extremes may require a 3rd bay
Omaha, NE	P	3416+	VHF yagi; all local VHF chs
Omaha, NE	P	4225	UHF yagi; all local UHF chs
Nevada			
Las Vegas, NV	G	3016	8 chs available
New Mexico			
Albuquerque, NM	G	3016	7 chs available
New York			
Albany/Troy/Schenectady, NY	P	3016+	5 chs available
Albany/Troy/Schenectady, NY	P	1473	VHF yagi for ch 13
Buffalo, NY	P	3603+	6 chs available; may need amp
Buffalo, NY	P	4221	Buffalo chs w/rotor
New York, NY	G	3016	8 chs, inc WPIX, WWOR
Rochester, NY	G	3016	Rochester chs only; not Buffalo
Syracuse, NY	P	3016+	5 chs available; some amp may be needed
Syracuse, NY	P	4193	UHF yagi for ch 68
North Carolina			
Charlotte, NC	X		
Greenville, Spartanburg, SC/ Asheville, NC	X		

Market name	Antenna Mkt type	Vendor/model	By market reception comments and restrictions
Greensboro/Winston Salem High Point, NC	X		
New Bern/Washington/Greenville Morehead City, NC	X		
Raleigh-Durham, NC	X		
Ohio			
Cincinnati, OH	G	3016	Rec. Cincinnati chs only
Cleveland, OH	G	3016	Metro ok; issue for Parma, South
Columbus, OH	P	3603+	Colum. chs only; no Cincy, Dayton
Columbus, OH	P	4308	Some issue in center of metro
Dayton, OH	G	3016	Only Dayton chs; No Cincy, Colum.
Toledo, OH	G	3016	Toledo chs only; Detroit chs w/rotor
Youngstown, OH	X		
Oklahoma			
Oklahoma, OK	G	3016	Rec. 7 chs
Tulsa, OK	P	3016+	5 chs available
Tulsa, OK	P	1473	VHF yagi for ch 11 and maybe 8
Oregon			
Portland, OR	G	3016	Rec. 7 chs; some amp may be required
Pennsylvania			
Altoona, Johnstown/Clearfield, PA	X		
Lancaster/York/Harrisburg, PA	X		
Philadelphia, PA	G	3016	Rec. 9 Philly chs
Pittsburgh, PA	P	3016+	May need jointenna for 3016
Pittsburgh, PA	P	3603	VHF yagi for ch 4
Wilkes-Barre, PA	G	4221	5 chs available; all UHF
Rhode Island			
Providence, RI	X		
South Carolina			
Charleston, SC	G	3016	5 chs available
Columbia, SC	X		
South Dakota			
Sioux Falls, SD	S	SF-1320 O	Outdoor mount only
Tennessee			
Chattanooga, TN	G	3016	All local chs available
Nashville, TN	X		
Memphis, TN	G	3016	6 chs available
Texas			
Austin, TX	G	3016	Metro ok; issues to west
Dallas, TX	G	3016	7 chs available
El Paso, TX	G	3016	5 chs available
Houston, TX	G	3016	7 chs available
San Antonio, TX	P	3016+	6 chs available
San Antonio, TX	P	4193	UHF yagi for ch 34
Texarkana, TX/Shreveport, LA	G	3016	5 chs available
Waco, TX	P	3016+	5 chs available
Waco, TX	P	4194	UHF yagi for ch 34

Complex "SMATV" distribution system

Market name	Antenna Mkt type	Vendor/model	By market reception comments and restrictions
Utah			
Salt Lake City, UT	G	3016	Rec. 6 chs
Vermont			
Burlington, VT	P	3016+	Extremes may need amp
Burlington, VT	P	3603	VHF yagi for ch 5
Virginia			
Norfolk/Hampton/Portsmouth, VA	G	3016	Rec. 6 chs
Richmond, VA	P	3016+	5 chs available
Richmond, VA	P	3603	VHF yagi for ch 6; may need jointenna
Roanoke/Lynchburg, VA	X		
Washington			
Seattle, WA	P	3016	May need jointenna for 3016; ch 13
Seattle, WA	P	1473	VHF yagi for ch 13
Seattle, WA	S	WS-1127	
Spokane, WA	P	3016+	5 chs available
Spokane, WA	P	3603	VHF yagi for ch 4; may need jointenna
West Virginia			
Charleston, WV	G	3016	6 chs avail; Huntington issues
Charleston, WV	P	3016+	Covers areas between Huntington
Charleston, WV	P	1473	and Charleston
Wisconsin			
Green Bay, WI	G	3016	Some issue with De Pere
Madison, WI	G	3016	5 chs available
Milwaukee, WI	G	3016	8 chs available

■ **Table 5-3 Frequencies of commercial TV channels.**

	Channel number	Frequency band MHz	Picture carrier MHz
Low Band	2	54–60	55.25
	3	60–66	61.25
	4	66–72	67.25
	5	76–82	77.25
	6	82–88	83.25
	FM	88–108	
Mid Band	A (14)	120–126	121.25
	B (15)	126–132	127.25
	C (16)	132–138	133.25
	D (17)	138–144	139.25
	E (18)	144–150	145.25
	F (19)	150–156	151.25
	G (20)	156–162	157.25
	H (21)	162–168	163.25
	I (22)	168–174	169.25
High Band	7	174–180	175.25
	8	180–186	181.25
	9	186–192	187.25
	10	192–198	193.25
	11	198–204	199.25
	12	204–210	205.25
	13	210–216	211.25
	J (23)	216–222	217.25
	K (24)	222–228	223.25

Commercial television channels

	Channel number	Frequency band MHz	Picture carrier MHz
Hyper Band	II (45)	348–354	349.25
	JJ (46)	354–360	355.25
	KK (47)	360–366	361.25
	LL (48)	366–372	367.25
	MM (49)	372–378	373.25
	NN (50)	378–384	379.25
	OO (51)	384–390	385.25
	PP (52)	390–396	391.25
	QQ (53)	396–402	397.25
	RR (54)	402–408	403.25
	SS (55)	408–414	409.25
	TT (56)	414–420	415.25
	UU (57)	420–426	421.25
	VV (58)	426–432	427.25
	WW (59)	432–440	435.25
	14	470–476	471.25
	15	476–482	477.25
	16	482–488	483.25
	17	488–494	489.25
	18	494–500	495.25
	19	500–506	501.25
	20	506–512	507.25
	21	512–518	513.25
	22	518–524	519.25

UHF Channel number	Frequency band MHz	Picture carrier MHz
43	644–650	645.25
44	650–656	651.25
45	556–662	657.25
46	662–668	663.25
47	668–674	669.25
48	674–680	675.25
49	680–686	681.25
50	686–692	687.25
51	692–698	693.25
52	698–704	699.25
53	704–710	705.25
54	710–716	711.25
55	716–722	717.25
56	722–728	723.25
57	728–734	729.25
58	734–740	735.25
59	740–746	741.25
60	746–752	747.25
61	752–758	753.25
62	758–764	759.25
63	764–770	765.25
64	770–776	771.25
65	776–782	777.25
66	782–788	783.25

		Frequency	Picture Carrier
L	(25)	228–234	229.25
M	(26)	234–240	235.25
N	(27)	240–246	241.25
O	(28)	246–252	247.25
P	(29)	252–258	253.25
Q	(30)	258–264	259.25
R	(31)	264–270	265.25
S	(32)	270–276	271.25
T	(33)	276–282	277.25
U	(34)	282–288	283.25
V	(35)	288–294	289.25
W	(36)	294–300	295.25

Super Band

		Frequency	Picture Carrier
AA	(37)	300–306	301.25
BB	(38)	306–312	307.25
CC	(39)	312–318	313.25
DD	(40)	318–324	319.25
EE	(41)	324–330	325.25
FF	(42)	330–336	331.25
GG	(43)	336–342	337.25
HH	(44)	342–348	343.25

Hyper Band

UHF

Channel	Frequency	Picture Carrier
23	524–530	525.25
24	530–536	531.25
25	536–542	537.25
26	542–548	543.25
27	548–554	549.25
28	554–560	555.25
29	560–566	561.25
30	566–572	567.25
31	572–578	573.25
32	578–584	579.25
33	584–590	585.25
34	590–596	591.25
35	596–602	597.25
36	602–608	603.25
37	608–614	609.25
38	614–620	615.25
39	620–626	621.25
40	626–632	627.25
41	632–638	633.25
42	638–644	639.25
67	788–794	789.25
68	794–800	795.25
69	800–806	801.25

Picture Carrier = Lower Frequency + 1.25 MHz
CATV channels shown in parentheses.
Color Carrier = Picture Carrier + 3.579 MHz
Sound Carrier = Picture Carrier + 4.5 MHz

115

Off-air antenna reference information

The big picture with the DSS system

IN THIS CHAPTER WE WILL SHOW YOU THE TOTAL DSS operation. This will be an overview of what you, the TV viewer, can expect from this marvelous entertainment system.

This chapter continues with the "big picture" of the total DSS system, system components, remote control functions, DSS menu flowcharts, Attraction menu, and many more functions.

Using the DSS system with multiple TV sets

To deliver its exceptional performance and benefits, the RCA brand DSS digital satellite system includes: an RCA satellite dish antenna with a single TV output connection, an RCA satellite receiver, and an RCA remote control.

If a consumer wishes to have the flexibility to control a second TV independently, the RCA DSS deluxe package provides that option. This package includes an RCA satellite dish antenna with two output connections, a deluxe RCA satellite receiver, and a universal remote that controls the RCA receiver, as well as multiple brands of televisions, VCRs, laserdisc players, and cable boxes.

This option gives consumers full performance, plus the capability to directly hook up two TVs that can be controlled independently. In addition, the dual-output capability allows the consumer to connect more than two TVs using a multiswitch (RCA D62041FD) in conjunction with signal diplexers (RCA D4001IFD). For each additional TV, the consumer will also need an additional RCA DSS satellite receiver.

Enhanced big-screen TV home theater

The RCA DSS digital satellite system raises the performance quality and entertainment options of television to a new level.

☐ Home theater technology exploits the high-resolution picture quality from the DSS digital satellite system.

☐ Home theater sound systems capture the CD-sound quality of the digital transmission.

☐ A wide variety of RCA home theater components let consumers "build" a system to match their lifestyle and budget.

State-of-the-art picture and sound

☐ The DSS system is capable of a clearer, sharper, ghost-free, laserdisc-quality picture.

☐ One-third more lines of horizontal resolution are available via DSS than in ordinary broadcast television or cable TV.

☐ The RCA DSS digital satellite system achieves this new level of performance by using powerful Hughes satellites and advanced digital video compression technology.

☐ Picture and sound are transmitted free of distortion and interference.

☐ Consumers can take advantage of ancillary home theater components, VCRs, and home theater audio systems.

State-of-the-art RCA DSS

The RCA DSS digital satellite system raises the television experience to a whole new level. Viewers can now enjoy digital reception capable of a clearer, sharper, ghost-free, laserdisc-quality picture, a picture with nearly one-third more lines of horizontal resolution than ordinary broadcast television or cable TV.

The RCA DSS receiving equipment delivers audio as impressive as that of a compact disc, providing unsurpassed performance for stereo and surround-sound systems.

The RCA DSS receiving equipment achieves this new level of performance by using powerful Hughes satellite and advanced digital video compression technology to send and receive signals. Picture and sound are transmitted in their purest form, free of the distortion and interference normally associated with current broadcast systems. This lets consumers take advantage of ancillary RCA home theater component systems.

The DSS digital satellite system uses digital video compression, which makes it possible to transmit signals on a single satellite with image quality superior to what is delivered by today's over-the-air broadcast and cable systems.

Programming is compatible not only with standard television, but with the new 16 × 9 TV format as well; and RCA DSS receiving equipment is already compatible with HDTV, a home entertainment breakthrough that is destined to become the new standard in American homes.

The big picture

Figure 6-1 shows the complete DSS satellite system. The satellite relays the programming signals back to your satellite dish. The satellite is located above the equator, in geostationary orbit approximately 22,300 miles above the earth. The uplink center transmits the programming material up to the satellite. The dish antenna receives the satellite signals from the "high-power bird."

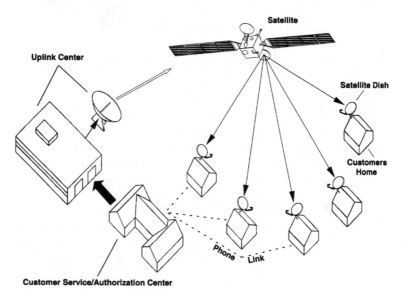

■ **6-1** *The complete DDS satellite system.* Thomson Consumer Electronics

The customer service/authorization center activates the subscription programming and process billing statements. Your DSS system is linked to the customer service center via the phone jack on the back of the DSS receiver.

DSS system components

Figure 6-2 shows how all of the DSS system components fit together in your home.

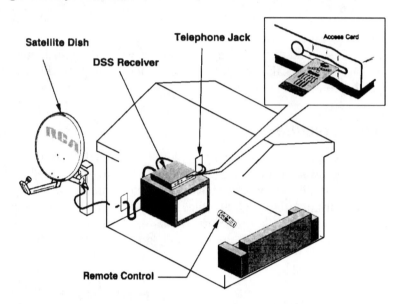

■ 6-2 *DSS system components for home reception.* Thomson Consumer Electronics

DSS receiver front panel

The front panel of the RCA DSS receiver is shown in Fig. 6-3 with all operation buttons and controls. These control buttons include on/off message, left/right and up/down arrows, the menu button, the select/display button, and the access card slot.

Remote control receiver functions

Figure 6-4 shows the DSS receiver remote control unit with locations and function readouts. The remote functions include DSS/VCR, on/off, ALT, guide, arrows, clear, TV/DSS, channel, volume, Mute, and many more.

DSS receiver back panel view

A view of the DSS receiver back panel and connection locations is shown in Fig. 6-5 on the following page. Some of these include the phone jack, low-speed data port, audio R and audio L, wideband data port, satellite signal in, and signal in from antenna and video player.

ON/OFF/MESSAGE: Turns the DSS receiver on and off. When the DSS receiver is turned off, a flashing light indicates that a message has been sent by the Customer Service Center.

TV/DSS: Switches between the satellite signal and the terrestrial antenna (or cable) when the DSS receiver is turned on.

ARROWS: The left/right & up/down arrows are used to point to different items in the menu system and program guide. May also be used to scroll up & down through the channels while viewing video.

MENU: Brings up the Main menu. If you are already in a menu, pressing MENU has the same effect as pressing SELECT.

ACCESS CARD SLOT: Insert your access card with the arrow face up and pointing toward the receiver. The DSS receiver is shipped with the access card inserted into the slot.

SELECT/DISPLAY: Selects items in the Program Guide and the menu system. When viewing a program, brings up a channel marker or header providing detailed program information. When previewing a coming attraction or pay-per-view event, this button provides purchase information.

■ **6-3** *DSS receiver front panel buttons/controls.* Thomson Consumer Electronics

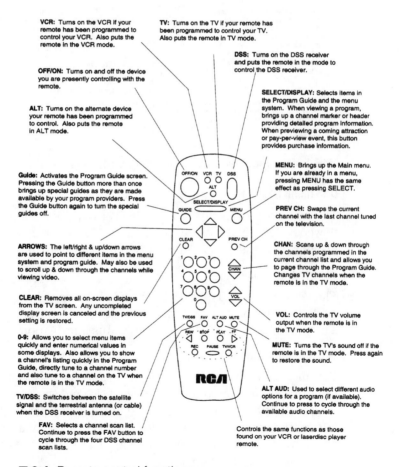

VCR: Turns on the VCR if your remote has been programmed to control your VCR. Also puts the remote in the VCR mode.

TV: Turns on the TV if your remote has been programmed to control your TV. Also puts the remote in TV mode.

DSS: Turns on the DSS receiver and puts the remote in the mode to control the DSS receiver.

OFF/ON: Turns on and off the device you are presently controlling with the remote.

ALT: Turns on the alternate device your remote has been programmed to control. Also puts the remote in ALT mode.

SELECT/DISPLAY: Selects items in the Program Guide and the menu system. When viewing a program, brings up a channel marker or header providing detailed program information. When previewing a coming attraction or pay-per-view event, this button provides purchase information.

Guide: Activates the Program Guide screen. Pressing the Guide button more than once brings up special guides as they are made available by your program providers. Press the Guide button again to turn the special guides off.

MENU: Brings up the Main menu. If you are already in a menu, pressing MENU has the same effect as pressing SELECT.

PREV CH: Swaps the current channel with the last channel tuned on the television.

ARROWS: The left/right & up/down arrows are used to point to different items in the menu system and program guide. May also be used to scroll up & down through the channels while viewing video.

CHAN: Scans up & down through the channels programmed in the current channel list and allows you to page through the Program Guide. Changes TV channels when the remote is in the TV mode.

CLEAR: Removes all on-screen displays from the TV screen. Any uncompleted display screen is canceled and the previous setting is restored.

VOL: Controls the TV volume output when the remote is in the TV mode.

0-9: Allows you to select menu items quickly and enter numerical values in some displays. Also allows you to show a channel's listing quickly in the Program Guide, directly tune to a channel number and also tune to a channel on the TV when the remote is in the TV mode.

MUTE: Turns the TV's sound off if the remote is in the TV mode. Press again to restore the sound.

TV/DSS: Switches between the satellite signal and the terrestrial antenna (or cable) when the DSS receiver is turned on.

ALT AUD: Used to select different audio options for a program (if available). Continue to press to cycle through the available audio channels.

FAV: Selects a channel scan list. Continue to press the FAV button to cycle through the four DSS channel scan lists.

Controls the same functions as those found on your VCR or laserdisc player remote.

■ **6-4** *Remote control functions.* Thomson Consumer Electronics

121

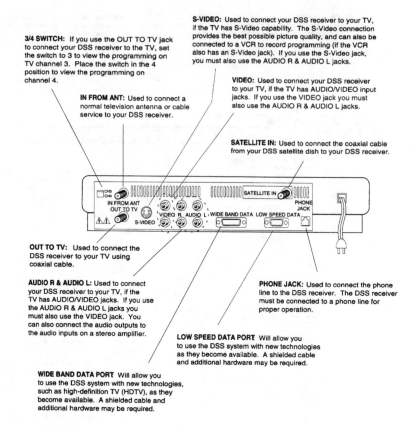

3/4 SWITCH: If you use the OUT TO TV jack to connect your DSS receiver to the TV, set the switch to 3 to view the programming on TV channel 3. Place the switch in the 4 position to view the programming on channel 4.

S-VIDEO: Used to connect your DSS receiver to your TV, if the TV has S-Video capability. The S-Video connection provides the best possible picture quality, and can also be connected to a VCR to record programming (if the VCR also has an S-Video jack). If you use the S-Video jack, you must also use the AUDIO R & AUDIO L jacks.

IN FROM ANT: Used to connect a normal television antenna or cable service to your DSS receiver.

VIDEO: Used to connect your DSS receiver to your TV, if the TV has AUDIO/VIDEO input jacks. If you use the VIDEO jack you must also use the AUDIO R & AUDIO L jacks.

SATELLITE IN: Used to connect the coaxial cable from your DSS satellite dish to your DSS receiver.

OUT TO TV: Used to connect the DSS receiver to your TV using coaxial cable.

AUDIO R & AUDIO L: Used to connect your DSS receiver to your TV, if the TV has AUDIO/VIDEO jacks. If you use the AUDIO R & AUDIO L jacks you must also use the VIDEO jack. You can also connect the audio outputs to the audio inputs on a stereo amplifier.

PHONE JACK: Used to connect the phone line to the DSS receiver. The DSS receiver must be connected to a phone line for proper operation.

LOW SPEED DATA PORT Will allow you to use the DSS system with new technologies as they become available. A shielded cable and additional hardware may be required.

WIDE BAND DATA PORT Will allow you to use the DSS system with new technologies, such as high-definition TV (HDTV), as they become available. A shielded cable and additional hardware may be required.

■ **6-5** *DSS receiver back panel connections.* Thomson Consumer Electronics

DSS menu flowchart

The flowchart for using the DSS receiver is shown in Fig. 6-6. The flowchart options include program guide, coming attractions, mailbox, Options menu, Alternate Audio, and Help menu.

The Program Guide

The Program Guide is shown on the following page in Fig. 6-7. The guide includes program information, the selection guide, selecting a theme for the guide, and program status information.

The Attractions menu

This menu gives you a list of the coming attractions that you may want to view. It gives you the program title, and the day, date, and time of the event. These listings also include pay-per-view buying information, purchase offers, additional viewing time information, and purchase cancelling information. The Attractions menu is shown on the following page in Fig. 6-8.

DSS Main Menu

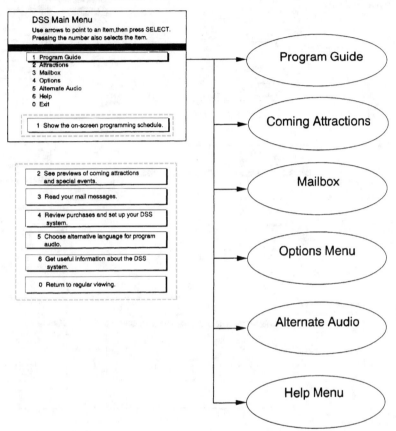

DSS Main Menu
Use arrows to point to an item, then press SELECT.
Pressing the number also selects the item.

1 Program Guide
2 Attractions
3 Mailbox
4 Options
5 Alternate Audio
6 Help
0 Exit

1 Show the on-screen programming schedule.

2 See previews of coming attractions and special events.

3 Read your mail messages.

4 Review purchases and set up your DSS system.

5 Choose alternative language for program audio.

6 Get useful information about the DSS system.

0 Return to regular viewing.

Program Guide

Coming Attractions

Mailbox

Options Menu

Alternate Audio

123

Help Menu

■ **6-6** *DSS menu flowchart.* Thomson Consumer Electronics

Mailbox and Alternate Audio menu

Refer to Fig. 6-9 on the following page for these menu features. The Mailbox menu lets you know if you have any mail messages and lets you exit, erase, and obtain help!

The Audio Alternate menu lets you select another audio language if you wish.

Options menu

The Options menu, shown on the following page in Fig. 6-10, lets you review or cancel upcoming programs, lock your system, set viewing limits, build your channel list, choose a channel list, set up your DSS system, and return to the main menu. Refer to Fig. 6-13 on the following page for Locks, Limits, and a Channel List menu.

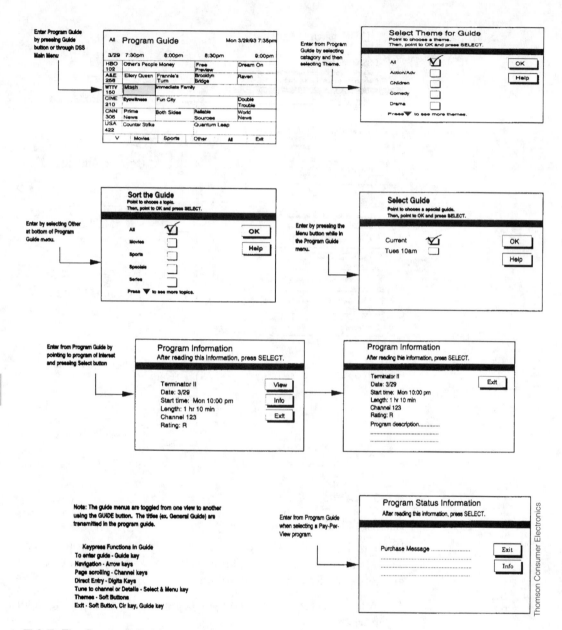

■ **6-7** *The Program Guide menu.*

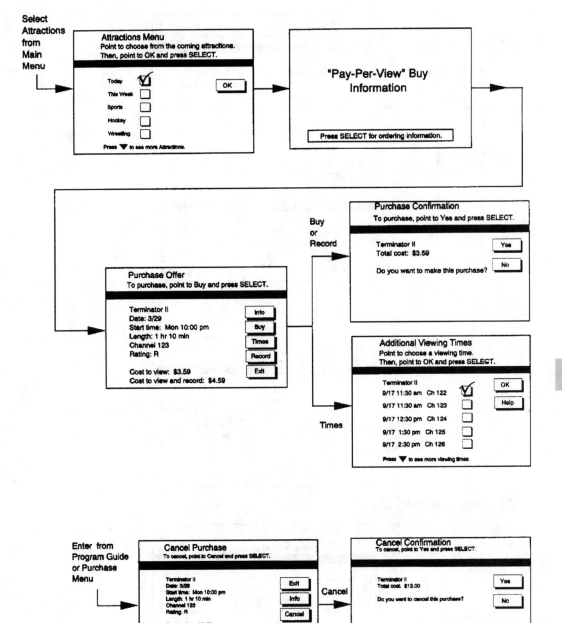

Select
Attractions
from
Main
Menu

Attractions Menu
Point to choose from the coming attractions.
Then, point to OK and press SELECT.

Today ☑
This Week ☐
Sports ☐
Hockey ☐
Wrestling ☐

OK

Press ▼ to see more Attractions.

**"Pay-Per-View" Buy
Information**

Press SELECT for ordering information.

Purchase Confirmation
To purchase, point to Yes and press SELECT.

Terminator II
Total cost: $3.59

Do you want to make this purchase?

Yes
No

Purchase Offer
To purchase, point to Buy and press SELECT.

Terminator II
Date: 3/29
Start time: Mon 10:00 pm
Length: 1 hr 10 min
Channel 123
Rating: R

Cost to view: $3.59
Cost to view and record: $4.59

Info
Buy
Times
Record
Exit

Buy
or
Record

Additional Viewing Times
Point to choose a viewing time.
Then, point to OK and press SELECT.

Terminator II
9/17 11:30 am Ch 122 ☑
9/17 11:30 am Ch 123 ☐
9/17 12:30 pm Ch 124 ☐
9/17 1:30 pm Ch 125 ☐
9/17 2:30 pm Ch 126 ☐

OK
Help

Press ▼ to see more viewing times.

Times

Enter from
Program Guide
or Purchase
Menu

Cancel Purchase
To cancel, point to Cancel and press SELECT.

Terminator II
Date: 3/29
Start time: Mon 10:00 pm
Length: 1 hr 10 min
Channel 123
Rating: R

Cost to view: $3.59

Exit
Info
Cancel
Help

Cancel

Cancel Confirmation
To cancel, point to Yes and press SELECT.

Terminator II
Total cost: $12.00

Do you want to cancel this purchase?

Yes
No

■ **6-8** *The Attractions menu.* Thomson Consumer Electronics

Mailbox Menu

From Main Menu

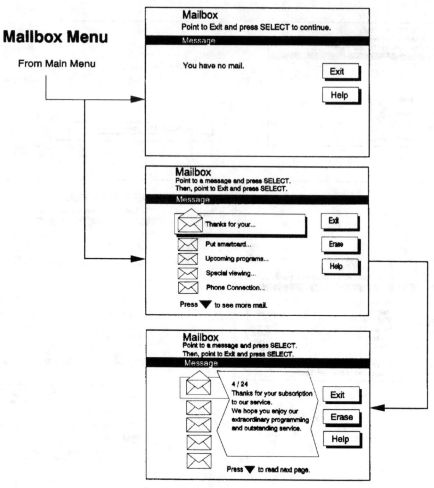

Alternate Audio Menu

Enter From the
Main Menu

Note: This menu only selects
the desired language if available.
This operation does not change
the written language of the menu.

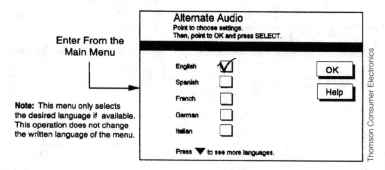

Thomson Consumer Electronics

■ **6-9** *The Mailbox and Alternate Audio menu.*

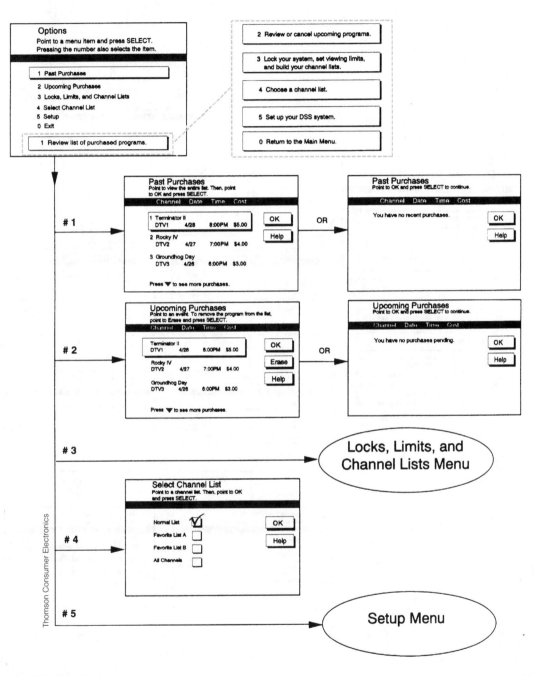

Options
Point to a menu item and press SELECT.
Pressing the number also selects the item.

1 Past Purchases

2 Upcoming Purchases
3 Locks, Limits, and Channel Lists
4 Select Channel List
5 Setup
0 Exit

1 Review list of purchased programs.

2 Review or cancel upcoming programs.

3 Lock your system, set viewing limits, and build your channel lists.

4 Choose a channel list.

5 Set up your DSS system.

0 Return to the Main Menu.

1

Past Purchases
Point to view the entire list. Then, point to OK and press SELECT.
Channel   Date   Time   Cost

1 Terminator II
DTV1   4/28   8:00PM $5.00

2 Rocky IV
DTV2   4/27   7:00PM $4.00

3 Groundhog Day
DTV3   4/26   6:00PM $3.00

Press ▼ to see more purchases.

OK

Help

OR

Past Purchases
Point to OK and press SELECT to continue.
Channel   Date   Time   Cost

You have no recent purchases.

OK

Help

2

Upcoming Purchases
Point to an event. To remove the program from the list, point to Erase and press SELECT.
Channel   Date   Time   Cost

Terminator II
DTV1   4/28   8:00PM $5.00

Rocky IV
DTV2   4/27   7:00PM $4.00

Groundhog Day
DTV3   4/26   6:00PM $3.00

Press ▼ to see more purchases.

OK

Erase

Help

OR

Upcoming Purchases
Point to OK and press SELECT to continue.
Channel   Date   Time   Cost

You have no purchases pending.

OK

Help

127

3

Locks, Limits, and
Channel Lists Menu

4

Select Channel List
Point to a channel list. Then, point to OK and press SELECT.

Normal List   ☑
Favorite List A ☐
Favorite List B ☐
All Channels   ☐

OK

Help

5

Setup Menu

Thomson Consumer Electronics

■ **6-10** *The Options menu.*

Setup menu

With the Setup menu you may select picture size, system test, system test results, and new access card data. Refer to Fig. 6-11.

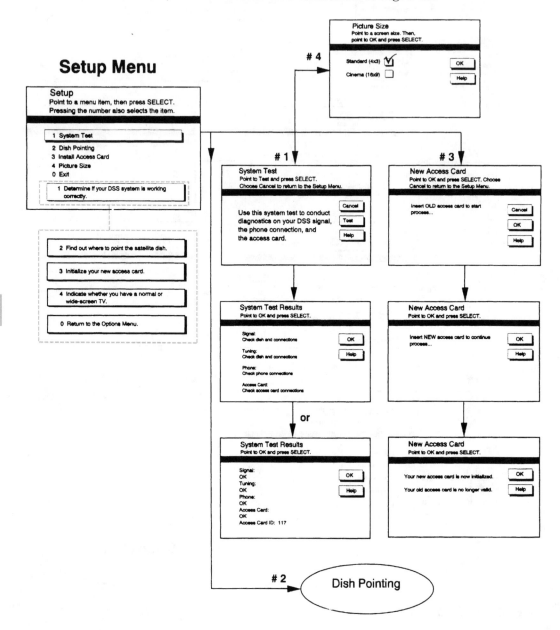

■ **6-11** *The Setup menu.* Thomson Consumer Electronics

Dish Pointing menu

With the Dish Pointing menu shown in Fig. 6-12 you can determine where to point the satellite dish, see and hear changes in signal strength, and then return to the Setup menu. You can bring up the Dish Pointing menu, Zip Code menu, Latitude and Longitude, and the Signal Meter.

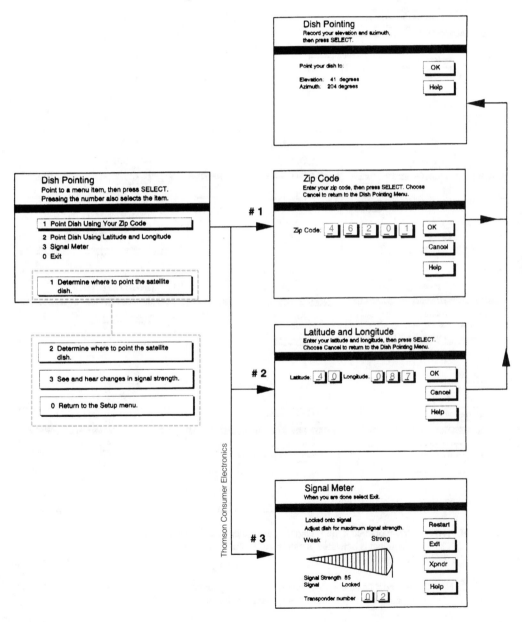

■ **6-12** *The Dish Pointing menu.*

Locks, Limits, and Channel Lists menu

This menu, shown in Fig. 6-13, lets you set ratings limits, spending limits, channel limits and lists, and enforce channel and spending limits.

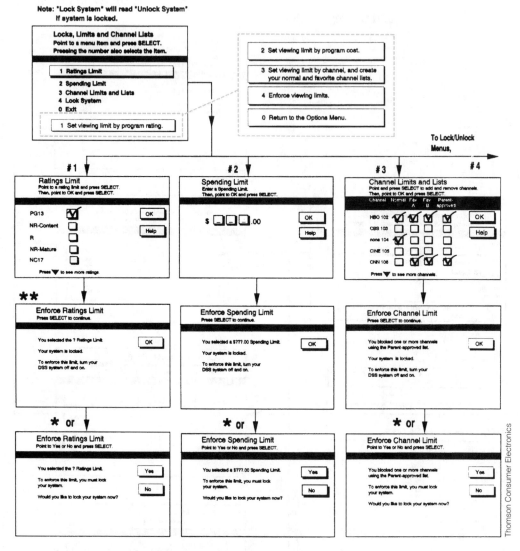

* **NOTE:** When enforcing the Ratings, Spending or Channel Limits, the menu that is activated is determined whether the system is "Locked" or "Unlocked".

** **NOTE:** "Ratings" limits may only be enforced if the your program provider supplies this capability in their program material.

■ **6-13** *The Locks, Limits, and Channel Lists menu.*

Thomson Consumer Electronics

NOTE: When enforcing the ratings, spending, or channel limits, the menu that is activated is determined by whether the system is locked or unlocked. Ratings limits may only be enforced if your program provider supplies this capability in their program material.

More Locks, Limits, and Channel Lists

Refer to Fig. 6-14 for the second channel Locks and Limits menu. These include lock confirmation. Notice that keys did not match,

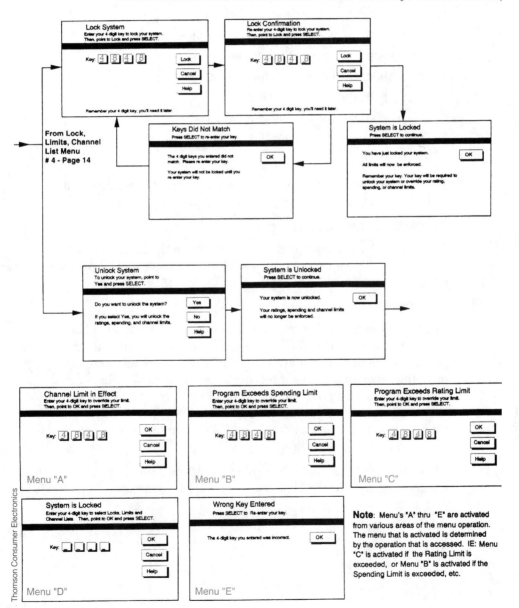

■ **6-14** *More Locks, Limits, and Channel Lists.*

that the system is locked, the channel limit is in effect, the spending limit is exceeded, and the wrong key entered.

Help menu

The Help menu (the menu we all need) is shown in Fig. 6-15. This menu brings up remote control help, program guide help, menu system help and program types. When in the Help menu you may select the following:

1. Learn about the remote control buttons.
2. Learn about using the on-screen program schedule.
3. Learn about the types of programs.
4. Learn about the On-Screen menu system.
5. Learn about the front panel buttons.
6. Learn about the back panel connections.
7. List of common DSS terms and definitions.
8. Return to the main menu.

The Help menu also features the remote control help, program guide help, menu system help, and the program types.

Channel Marker and More Help menu

Some more Help menu information is shown on the following page in Fig. 6-16. These Help menus contain the glossary, back panel help, and front panel help.

DSS system

Figure 6-17 on the following page shows the complete RCA DSS receiving system. The three essential components of the DSS (digital satellite system) are the 18-inch (pizza-sized) antenna dish, the receiver, and the interactive remote control, which are all manufactured under the RCA brand by Thomson Consumer Electronics. The DSS componentry is compatible with the digital programming made available by USSB (United States Satellite Broadcasting) and DirecTV.

All photos, drawings, and technical information in this chapter are used with permission of Thomson Consumer Electronics.

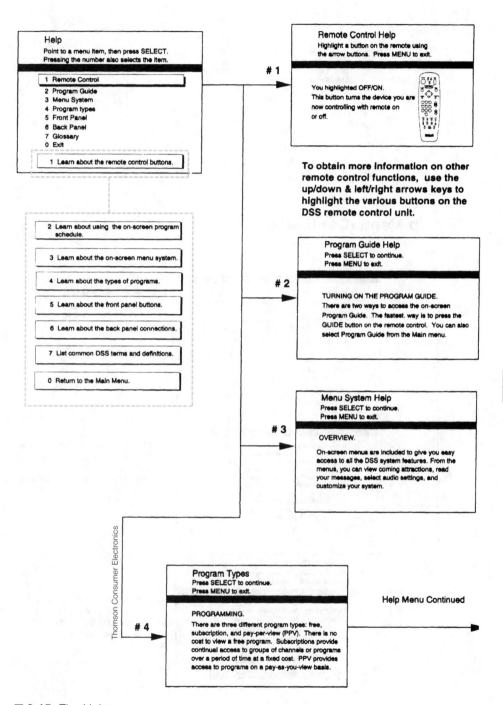

Help

Point to a menu item, then press SELECT.
Pressing the number also selects the item.

1 Remote Control
2 Program Guide
3 Menu System
4 Program types
5 Front Panel
6 Back Panel
7 Glossary
0 Exit

1 Learn about the remote control buttons.

2 Learn about using the on-screen program schedule.

3 Learn about the on-screen menu system.

4 Learn about the types of programs.

5 Learn about the front panel buttons.

6 Learn about the back panel connections.

7 List common DSS terms and definitions.

0 Return to the Main Menu.

 # 1

Remote Control Help

Highlight a button on the remote using
the arrow buttons. Press MENU to exit.

You highlighted OFF/ON.
This button turns the device you are
now controlling with remote on
or off.

To obtain more information on other remote control functions, use the up/down & left/right arrows keys to highlight the various buttons on the DSS remote control unit.

 # 2

Program Guide Help

Press SELECT to continue.
Press MENU to exit.

TURNING ON THE PROGRAM GUIDE.
There are two ways to access the on-screen
Program Guide. The fastest way is to press the
GUIDE button on the remote control. You can also
select Program Guide from the Main menu.

 # 3

Menu System Help

Press SELECT to continue.
Press MENU to exit.

OVERVIEW.

On-screen menus are included to give you easy
access to all the DSS system features. From the
menus, you can view coming attractions, read
your messages, select audio settings, and
customize your system.

Thomson Consumer Electronics

 # 4

Program Types

Press SELECT to continue.
Press MENU to exit.

PROGRAMMING.

There are three different program types: free,
subscription, and pay-per-view (PPV). There is no
cost to view a free program. Subscriptions provide
continual access to groups of channels or programs
over a period of time at a fixed cost. PPV provides
access to programs on a pay-as-you-view basis.

Help Menu Continued

■ **6-15** *The Help menu.*

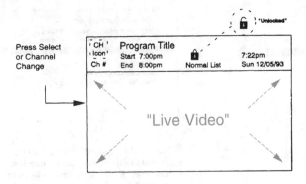

Press Select
or Channel
Change

CH   Program Title
Icon   Start 7:00pm   7:22pm
Ch #   End 8:00pm   Normal List   Sun 12/05/93

"Unlocked"

"Live Video"

Help Menu (Continued)

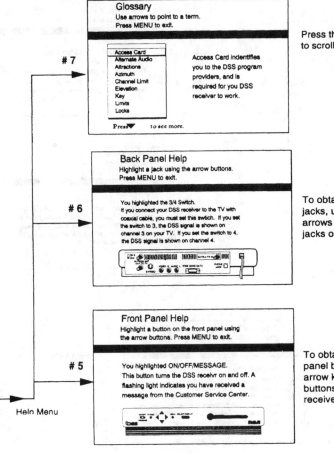

Glossary
Use arrows to point to a term.
Press MENU to exit.

7

Access Card
Alternate Audio
Attractions
Azimuth
Channel Limit
Elevation
Key
Limits
Locks

Access Card identifies
you to the DSS program
providers, and is
required for you DSS
receiver to work.

Press ▼ to see more.

Press the up/down arrow key
to scroll through the list.

Back Panel Help
Highlight a jack using the arrow buttons.
Press MENU to exit.

6

You highlighted the 3/4 Switch.
If you connect your DSS receiver to the TV with
coaxial cable, you must set this switch. If you set
the switch to 3, the DSS signal is shown on
channel 3 on your TV. If you set the switch to 4,
the DSS signal is shown on channel 4.

To obtain information on the other
jacks, use the up/down & left/right
arrows keys to highlight the various
jacks on the rear of the DSS receiver.

Front Panel Help
Highlight a button on the front panel using
the arrow buttons. Press MENU to exit.

5

You highlighted ON/OFF/MESSAGE.
This button turns the DSS receivr on and off. A
flashing light indicates you have received a
message from the Customer Service Center.

To obtain information on other front
panel buttons, use the left/right
arrow keys to highlight the various
buttons on the front of the DSS
receiver.

Help Menu

■ **6-16** *The Channel Marker and Help menu.* Thomson Consumer Electronics

134

Thomson Consumer Electronics

■ **6-17** *The three essential components of the RCA DSS system.*

The DBS satellite system

IN THIS CHAPTER YOU WILL FIND OUT HOW THE HIGH-powered DBS satellites are operated, checked out, and launched into geosynchronous orbit 22,300 miles out into space. You will find out that launching these birds is a very tricky and risky business.

There will also be information on the DSS satellite built by GM Hughes Space Communications Company for Hughes Communications, Inc. These are all units of the GM Hughes Electronics System. The chapter closes with a brief history of various satellites that have been launched into orbit from various parts of the world.

Hughes DBS satellites

High-powered satellites built by Hughes Space and Communications Company (HSC) began bringing true direct TV broadcast satellite (DBS) service to homes throughout North America in 1994.

The spacecrafts are HS-601 body-stabilized models ordered by Hughes Communications, Inc. (HCI). A drawing of the HS-601 unfolded as in space operation is shown in Fig. 7-1 on the following page. The first bird is shared by DirecTV, Inc., a GM Hughes Electronics subsidiary that operates 11 transponders, and by United States Satellite Broadcasting, which operates 5 transponders. A second spacecraft is operated solely by DirecTV, to increase its channel capacity. Together the satellites are capable of delivering more than 150 channels of entertainment programming to subscribers using the small DSS dishes and the required in-home digital receiver decoders.

The photo in Fig. 7-2 on the following page shows the GM Hughes HS-601 body-stabilized spacecraft that is parked at 101 degrees west longitude. This satellite carries 16 transponders that each produce 120 watts of transmitter power. These are among the most powerful commercial satellites ever built and put into orbit. As noted previously, these HS-601 spacecraft are built by Hughes Space and Communications Company.

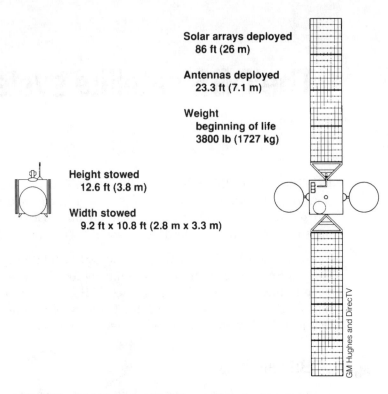

Solar arrays deployed
86 ft (26 m)

Antennas deployed
23.3 ft (7.1 m)

Weight
beginning of life
3800 lb (1727 kg)

Height stowed
12.6 ft (3.8 m)

Width stowed
9.2 ft x 10.8 ft (2.8 m x 3.3 m)

GM Hughes and DirecTV

■ **7-1** *The GM Hughes HS-601 satellite, deployed.*

To provide the high satellite power necessary for transmitting to such small receiving antennas, each DBS spacecraft has 16 120-watt traveling-wave tube amplifiers (TWTAs). The TWTAs can be reconfigured to provide eight channels, each with 240 watts of power. The amplifiers are suitable for analog or digital signals, and are capable of transmitting high-definition television (HDTV) signals and compact-disc-quality audio as well.

Engineers at Hughes Space and Communications are shown on the following page in Fig. 7-3 checking out the antennas on DBS-1, North America's first direct-to-home television satellite. The two large antennas are for transmitting signals to Earth, while the smaller round dish in the center of the photo is the receiving antenna. Also visible is the complex wiring for the satellite's 16 powerful transponders.

The satellite operates in the BSS portion of the Ku-band spectrum (12.2 to 12.7 GHz) and employ circular polarization. Depending on the configuration of the transponder, they can deliver 48 to 53 dBW radiated power over the contiguous United States and southern Canada. The spacecraft will be collocated at 101 degrees west longitude.

GM Hughes and DirecTV

■ **7-2** *The GM Hughes HS-601 in the "box" above the equator.*

An Ariane 4 rocket carried DBS-1 into space. DBS-1 is used by DirecTV and USSB. As shown on the following page in Fig. 7-4, it was successfully launched on December 17, 1993 from Kourou, French Guiana. The Hughes HS-601 model satellite is shown in this photo on its launch pad aboard an Ariane rocket.

An Atlas IIAS rocket boosted DBS-2 on August 3, 1994. A third satellite was launched in 1995 on an Ariane rocket. The booster takes the spacecraft to geosynchronous transfer orbit. The spacecraft's integral liquid apogee motor raises it to geostationary orbit 22,300 miles (36,000 KM) above the equator. The flight-proven bipropellant propulsion system includes not only the 110-lb./f.

GM Hughes and DirecTV

■ **7-3** *Engineers at Hughes checking out antennas on DBS-1 satellite for DirecTV.*

Marquardt apogee motor, but also a dozen 5-lb./f. thrusters for station-keeping during each satellite's 12 years in orbit.

Each DBS spacecraft measures 23.3 feet (7.1 meters) across with the two transmit antennas deployed, and 86 feet (26 meters) long from the tip of one four-panel solar array wing to the other. These arrays generate a combined 4300 watts of electrical power, backed up by a 32-cell nickel-hydrogen battery for uninterrupted power during eclipse. The spacecraft weigh around 3800 pounds (1727 Kg) at the beginning of their life in orbit.

An innovative graphite antenna system makes its debut on the DBS spacecraft. The transmit and receive reflectors feature a specially contoured surface that requires only one, rather than multiple, feedhorns to provide an optimal signal. The composite material is so light that each 8-foot (2.4 meter)-diameter transmit antenna weighs less than 20 pounds (9.1 Kg). The antennas are aligned in a unified structure to provide a significant improvement in antenna-pointing performance.

The HS-601 satellite line was introduced by HSC in 1987 to meet anticipated requirements for high-power, multiple-payload spacecraft for such applications as direct broadcast, private business

GM Hughes and DirecTV

■ 7-4 *The launch of the Hughes DBS-1 satellite for DBS.*

networks, and mobile communications. By mid-1994, customers had ordered 35 HS-601 satellites in various configurations.

The HS-601 body is composed of two main modules. The bus module is the primary structure that carries the launch vehicle loads and contains the propulsion, attitude control, and electrical power subsystems. The payload module is a honeycomb structure that contains the payload electronics, telemetry, command and ranging equipment, and the isothermal heat pipes. Reflectors, antenna feeds, and solar arrays mount directly to the primary module, and antenna configurations can be placed on three faces of the bus.

Such a modular approach allows work to proceed in parallel, thereby shortening the manufacturing schedule and test time. DBS-2 is shown being prepared for launch in Fig. 7-5.

GM Hughes and DirecTV

■ **7-5** *Hughes engineers making the final check-out on the DBS-2 satellite.*

The operations control center for the DBS satellites is at HCI headquarters in El Segundo, CA; telemetry and command terminals are in Castle Rock, CO, and Spring Creek, N.Y. Uplink is from the DirecTV Castle Rock Broadcast Center, which is capable of transmitting 216 simultaneous broadcast channels to the satellites.

Hughes spacecraft launch information

From a news release dated December 13, 1993:

A pair of "firsts" is riding on the next Ariane 4 rocket launch, in the form of two satellites built by Hughes Aircraft Company.

Copassengers for Friday's (Dec. 17) event are DBS-1, which will bring the first direct-to-home television service to the United States and Canada; and Thaicom 1, the first of two spacecraft for Thailand's national satellite system. Launch is set for 10:27 p.m. local time from the Guiana Space Center in Kourou.

This is the second time two Hughes-built satellites will be sharing an Arianespace rocket. The previous mission was the successful launch of SBS-6 and Galaxy VI on an Ariane 4 in October 1990.

DBS-1 will leave the rocket first, about 21 minutes after liftoff. This is one of the most powerful commercial satellites ever launched. It is a Hughes HS-601 model that will be operated by DirecTV, Inc. and United States Satellite Broadcasting. The signal from this high powered bird will be picked up by the small DSS dishes all across the United States and southern Canada. It carries 16 transponders, each with 120 watts of RF power. DirecTV will operate 11 of the 16 transponders and USSB will have the other five. A twin of this DBS satellite was launched in 1995, and with digital compression, the satellite will be able to transmit up to 150 channels of programming.

Thaicom 1 is an HS 376L model satellite, tailored to the specific telecommunications needs of Thailand. It is due to separate from the Ariane about 25 minutes after launch. This is a new, lightweight version of Hughes workhorse HS 376 model, with 10 transponders in C-band and two in Ku band (HS 376 models can carry up to 24 transponders). Shinawatra Satellite Public Co. Ltd. (SSA) will operate this satellite and THAICOM 2, which is set for launch in 1995.

Hughes engineers in the Mission Control Center in El Segundo, CA., will "fly" both satellites after their separation from the Ariane rocket. For the DBS mission, Hughes controllers will also use ground stations in Sydney, Australia, and Castle Rock, CO. For THAICOM, stations in Cibinong, Indonesia, and Fillmore, CA, will be used, and a Thai controller and two Thia orbital analysis will be on the team.

SSA is a subsidiary of Shinawatra Computer and Communications Co. (SC&C), designated to operate the first satellite communications system in Thailand.

On Sept. 11, 1991, SC&C was entrusted and awarded the 30-year concession from the government of Thailand to acquire, launch and operate the satellite project. In November 1991, SSA was thus founded to carry out the project under the government's supervision and promotion.

His Majesty King Bhumiphol officially named the satellite "Thaicom," symbolizing the link between Thailand and modern communications technology.

Not only will these satellites that are launched introduce new services, but they also are milestones for Hughes. DBS-1 and Thaicom 1 will be, respectively, the 100th and 101st Hughes-designed communications satellites to be launched. Hughes started the geosynchronous communications satellite business 30 years ago, with the launch of Syncom. That 78-pound spacecraft was the first satellite that could be used 24 hours a day, because it was the first put into geosynchronous orbit 22,300 miles above the earth. In that orbit, the satellite circles the planet at the same speed that Earth rotates, making the spacecraft seem to be stationary over a point on Earth.

The new satellites will fly only a month after another Hughes-Ariane mission. The Solidaridad 1 spacecraft, built for Mexico, was launched November 19 and is now undergoing in-orbit tests. Altogether, in 1993, Arianespace will have launched five Hughes-built satellites. Next year's schedule is even more ambitious. Out of the 15 Hughes spacecraft slated for launch, eight will be on Ariane rockets.

The spacecraft are manufactured in the El Segundo facilities of Hughes Space and Communications Company (HSC), the world's leading producer of commercial communications satellites. Hughes has built more than 40 percent of those now in service. HSC and DirecTV are part of the Hughes Aircraft Company, which in turn is a unit of GM Hughes Electronics.

Hughes triple play launch

A news release, dated July 15, 1994, stated that Hughes would launch three satellites from three sites around the globe. Crews from Hughes Space and Communications Company (HSC) were in place in the United States, French Guiana, and China, preparing three satellites for launch in a 10-day period.

All three were communications satellites, built near Los Angeles by HSC to bring television, telephone, direct broadcast TV, and other services to users on three continents. Together with two other satellites, this will make five Hughes satellites orbited in a little over one month for the GM Hughes Electronics Company.

This was Hughes' busiest launch year until that time, with 13 spacecraft slated for launch through December, and 14 in 1995. The Galaxy I-R satellite went into service only weeks after its February launch. Hughes is the world's leading manufacturer of commercial communications satellites, and a major supplier to the U.S. government. Since it built and orbited its first satellite (Syncom) 31 years ago, Hughes has launched 104 communications satellites and has accumulated enormous experience in space operations.

Intelsat satellite launch

Many readers may find launching of these "big birds" and going through the countdown at Cape Canaveral quite interesting. Let's look at one of these Lockheed Martin launches of an INTELSAT satellite in the first part of 1995. Destined for an orbital location a 310 degrees east, INTELSAT 705 had a good liftoff from Launch pad 36 at Cap Canaveral. Fla. This launch put the INTELSAT series 7 into its home spot for 14 years, 22,000 nautical miles out in space.

Launching a satellite is not your normal day-at-the-office business. Things are not always perfect in the days before the launch. If you have followed any of this industry's news, you must recall that there have been many launch failures in the past few months. This scheduled launch started out a couple days late.

As most of you who keep up with the moon and space shuttle launches know, the weather can be a problem at the Cape, and shots have to be rescheduled for the next window of opportunity. Many times it's a waiting game.

Ready for the launch

After some problems and weather delays, the Intelsat has been rescheduled for a Wednesday morning lift-off. The launch window is now open at 0118 hours. The Atlas/Centaur rocket is poised for a trip into space. The weather is very good and is rated as excellent. In fact, it is the best weather for a launch in many years.

The launch site is busy as bees on Tuesday evening at 2,000 hours as the mobile service tower starts its trip back from the launch vehicle in 15 or so minutes at about four miles per hour. Refer to Fig. 7-6 on the following page for the T-times of countdown to liftoff.

At 90 minutes before the scheduled liftoff, the fuel is loaded into the launch vehicle. Liquid oxygen is maintained at minus 290 degrees, and the liquid hydrogen near absolute zero. At T-minus 90 seconds, the fuel will be topped off at flight levels and secured.

145

T-Time (min:sec)	Event
T-163	Man stations for integrated launch operations
T-160	Start tower removal
T-157	**Range countdown starts**
T-125	Start C-band systems test
T-105	**T-105 minutes and holding for 30 minutes**
———	Weather briefing (15 min. prior to end of hold)
	Establish road blocks (15 min. prior to end of hold)
———	Ready reprot for cryogenic tanking (3 minutes prior to end of hold)
T-105	**Resume count**
T-104	Pressurize to Step II pressures
T-092	Start Centaur LO_2 tanking
T-077	Start LH_2 chilldown. Seal blockhouse doors.
T-075	Start Atlas LO_2 tanking
T-059	Start Centaur LH_2 tanking
T-032	Start flight termination system self-test
T-014	Final prelaunch weather briefing
T-005	**T-5 minutes and holding for 15 minutes**
	Status report for pickup of count
T-005	**Resume count**
T-002:15	Start flight pressurization
T-002	Centaur to internal power, flight termination system to ARM
T-000:31	Launch sequence start
T-000:01	Launch release

Mark Events

	T-Time	Event
0.	T-000:00	**Liftoff**
1.	T+59.4	Air-lit SRB ignition
2.	T+1.07.6	Jettison ground-lit SRBs
3.	T+1.54.8	Jettison air-lit SRBs
4.	T+2:44.7	Booster engine cutoff
5.	T+2.47.8	Jettison booster package
6.	T+3:30.1	Jettison payload fairing
7.	T+4:47.4	Sustainer engine cutoff
8.	T+4.49.4	Atlas-Centaur separation
9.	T+5.05.9	Centaur first main engine start
10.	T+09.45.9	Centaur main engine cutoff and begin Centaur coast
11.	T+24:30.4	Centaur second main engine start
12.	T+26:05.1	Centaur main engine cutoff
13.	T+28:16.1	**Spacecraft separation**

■ **7-6** *The countdown time frame for a satellite launch.* GM Hughes and DirecTV

A planned hold at T-minus five minutes is reached, and additional checks and safeguards are made by the mission controllers. Checks are made over the sound system with GOs from each controller. The mission controllers are responsible for certain systems of the launch vehicle. For this launch, the voice call check-outs on all of the status reports come back positive, and the five-minute planned hold is completed and normal countdown proceeds.

Thus the countdown is resumed. At the two-minute-and-counting point, the responses are getting faster and it's difficult to keep up with all the responses as they are announced. This launch is the latest of over 100 Centaur launches, so these folks really know what they are doing. Everything goes very smoothly, and with a

146

$100 million satellite and a $100 million launch, nobody is willing to allow much room for error.

The countdown is now zipping along, and the launch sequence is started manually. Its now time to push that "RED" button, which indicates the last chance of human error before the launch vehicle leaves the ground.

Launch time is reached at T-minus 1 second. Engine ignition occurs, and you sure feel and hear it. The launch vehicle now lifts off pad 36B on a one-way trip out into space. I am told the actual rocket liftoff is reported back to the control center by a two-inch movement when a wire connected at the bottom of the rocket pulls apart. So much for high-tech devices.

Control center progress reports

After the launch, the mission control center (Fig. 7-7) reports that all anticipated values are normal. The launch engineers are now reporting the various stages and progress of the launch in a very concise manner. In the control center there are lots of telemetry lab techs reading strips of papers generated by endless recorders.

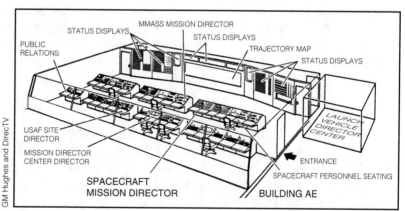

Cape Canaveral Mission Director's Center

■ **7-7** *The layout of the mission control center at Cape Canaveral.*

If you are at the launch center during this time, the launch control narrator for the Atlas makes every detail much more understandable from the layman's perspective. You will also have excellent coverage of the launch, coverage that is provided by color cameras whose signals are fed live to Galaxy 3 in Miramar, Florida. The pictures from the cameras following the launch vehicle are spectacular. A large piece of glass sits in front of the camera lens and produces quality pictures that the human eye cannot see. From this camera advantage point you can see the SRBs falling away as their fuel is spent. The satellite then goes into the super-synchronous transfer orbit, 20,769 nautical miles above the equator. Now, after reaching the intended orbit and expending all the fuel on the Centaur, spin-up of the satellite occurs and separation is accomplished. Another satellite launch mission success!

After separation has occurred, and spin-down is accomplished, the next occurrence is the first of the multiple apogee motor firings. Now the satellite will find its way to 310 degrees east longitude over the next few weeks. A "live vehicle" in this launch provides for a transfer at 21,810 nautical miles. The higher orbit will require less fuel expended by the apogee kick motors, and thus provide a longer life for the Intelsat 705 station, which normally is expected to last for 14 years.

More satellite launching info

Because of the many things that can go wrong, not to mention weather conditions, all rocket satellite launches are not successful. GM Hughes lost a communications satellite in the first part of 1995; Pan Am Sat lost PAS-3 in December of 1994, and AT&T lost Telstar 402 during the summer of 1994. These are just a few of the losses that will be felt in the world communications industry for many years. However, there is a greater-than-95% chance that a good satellite launch will occur.

The launch

Getting a 10,000-pound or so satellite into the proper orbit is no easy task. It takes less than 30 minutes to launch the rocket with spacecraft on top to the transfer orbit, at the point where the launch vehicle and spacecraft part ways. This procedure is then a very critical part of the satellite's life.

Let's now look at an Intelsat-type satellite launch on a Atlas IIAS rocket at the Cape by Martin Marietta. This satellite is to be put into a location of 310 degrees east, to be used to replace an older Series 5 satellite.

There are only five companies that launch commercial or civilian satellites worldwide. McDonnell Douglas (the Delta series) and Martin Marietta (the Atlas series) provide launch services in the United States. Other launch companies are Arianespace (the Ariane series in French Guiana), The Long March series for the Chinese, and the Pronton series for the Russians.

All of these organizations use their own launch buggies and methods to get these satellites into orbit, but we will now see how Martin Marietta goes about it at the Cape.

Their launch vehicle setup is as follows:

☐ A solid rocket booster stage powered by Thiokol Caster 1VA solid rocket boosters (SRB).

☐ A core vehicle stage (booster and sustainer) powered by Rockendyne MA-5A liquid propellant engines (RP-1 fuel and liquid oxygen).

☐ A Centaur upper stage powered by two Pratt and Whitney RL10A-4-1 liquid hydrogen and liquid oxygen engines with extendible nozzles.

☐ A 14-foot metal payload fairing that protects the spacecraft during ascent.

Martin Marietta has been launching from the Space and Launch Complex at Cape Canaveral Air Station, located east of Orlando, Florida. However, the West Coast launch complex will be ready in 1996, after its renovation is completed to accept the larger Atlas launch vehicles.

Launch overview

This will give you a brief overview of what happens during these satellite launches. The mission control center provides overall management control for the launch. The controllers responsible for the mission provide the needed go/no-go control and coordination. once all the checks have been verified and just before liftoff, the three main Atlas engines and two solid rocket boosters fire. Their performance is verified before Atlas is released from the launch complex.

When there is liftoff, Atlas begins its initial pitch-over and roll into a prescribed ascent profile to acquire the specified aerodynamic loads. Approximately one minute into the flight, the first two SRBs (solid rocket boosters) burn out and the second set of SRBs ignite. The first pair are then jettisoned when range safety conditions are met. The drawings in Figs. 7-8 and 7-9 on the following page will give you a better understanding of the total launch sequences.

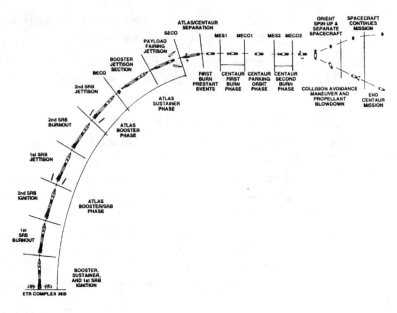

■ 7-8 *The call-out for various time frames of a satellite launch from Cape Canaveral.* GM Hughes and DirecTV

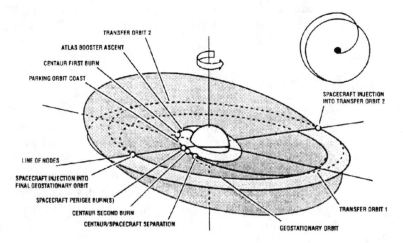

■ 7-9 *How the satellite goes in orbit and arrives in the "box."* GM Hughes and DirecTV

At about two minutes into the flight, the second set of SRBs are jettisoned as their fuel is expended. At approximately two-and-one-half minutes into the flight, booster engine cutoff (BECO) occurs, and the booster section is also jettisoned. Then, for the next two minutes, the Atlas is in the sustainer solo phase. After the next 30 seconds into the solo phase, the payload fairing is jettisoned.

In this mode, the propellant is being burned at a rate of 280 pounds per second. At this point the Atlas is at about 100 miles in altitude and 325 miles downrange, whipping along at close to 9,840 miles per hour. At this point, sustainer engine cutoff (SECO) occurs, and it is immediately followed by Atlas-Centaur separation.

The Pratt and Whitney RL-10 main engine nozzles are extended and main engine start (MES-1) now occurs. At this time we are about five minutes into the flight. The satellite and Centaur are now in "first burn," the longer of two Centaur firings. This first burn lasts about five minutes and will place the launch vehicle and payload in a slightly elliptical orbit for parking. Then, after MECO-1 (main engine cutoff), the Centaur and satellite will coast for about 13½ minutes.

One last blast

When the coordinated start position is reached (which is just north of the equator for launches from Cape Canaveral), the Centaur main engines are reignited for 1½ minutes, MES-2. The main engines will put the vehicle into the required geostationary transfer orbit. After the Centaur main engine's second burn, MECO-2 is now reached.

The spacecraft (satellite) is now separated from the launch vehicle. Just prior to separation, about four minutes after MECO-2, the Centaur provides the required spin-up and specific pointing attitude required to put the spacecraft into geostationary transfer orbit.

As a precaution for not contaminating the trajectory of the free-flying spacecraft, Centaur will de-spin and perform a series of maneuvers called *collision and contamination avoidance maneuvers*. When this sequence is completed, the Centaur mission is completed, and Centaur will unlock its vent valves and terminate its programmed flight plan.

Once the spacecraft has been successfully placed into Transfer Orbit 1 by the launch vehicle, the spacecraft's liquid apogee kick motors are used to enlarge the orbit. Enlarging the orbit takes about three to four days.

It now takes from three to four weeks for the satellite to drift east/west to its assigned orbital location. This is performed very slowly to avoid overshooting its parking spot and to conserve onboard fuel. Keep in mind that this onboard fuel is used throughout the satellite's service lifetime to keep it "on target" in its assigned "box."

The box is what allows you to have a stationary antenna at the ground receiving dish location. This keeps you from having to move the dish if the satellite would be drifting out of the box. At this time, most satellites are kept at least within 0.1 degrees, but some of the newer spacecraft are capable of 0.05 degrees of orbital stability, especially satellites with Ku-band transmission capability.

When a successful launch is completed, the satellite is put through a period of in-orbit testing. These tests may take several weeks, and some can be made while the satellite drifts to its final operational parking box spot. These test objectives include verification of deployment of arrays and antennas, verification of system operations, and pattern verification, which are more accurately determined in space. These tests not only confirm the operational aspects of the satellite, but can be used later on as a baseline or reference as the performance deteriorates over the life of the satellite.

Martin Marietta's launch of DBS-2

The second high-powered direct broadcast satellite (DBS), called DBS-2, was successfully launched by Martin Marietta on August 3, 1994 from Cape Canaveral Air Force Base in Florida. The photo of this launch is shown in Fig. 7-10. The Hughes HS-601 satellite was launched into space by a Atlas IIA rocket, and will be used by DirecTV to expand the programming lineup to 150 channels for the RCA brand DSS system.

Figure 7-11 on the following page shows the GM Hughes technicians adjusting the thermal blankets on the DBS-2 wide-coverage satellite. All technical information, drawings, charts, and photos in this chapter are courtesy of DirecTV, a division of GM Hughes Electronics.

GM Hughes and DirecTV

■ **7-10** *The launching of the GM Hughes DBS-2 satellite by Martin Marietta, August 1994.*

GM Hughes and DirecTV

■ **7-11** *Hughes technicians adjusting the thermal blankets on the DBS-2 satellite to be used by DirecTV. The blankets ensure that the satellite maintains a constant internal temperature in orbit.*

What you can watch with the DSS dish

8

IN THIS CHAPTER YOU WILL FIND OUT WHAT PROGRAM services are available with the Digital Satellite Service, and we will answer the question "What can I watch from these high-powered birds"?

In this chapter you will find USSB, DirecTV, and Hubbards DBS programming and viewing information, the cost of this service for your viewing pleasure, a list of the programs and services that you will be able to view at the time of this book's publication, and a look into the crystal ball of what programs and features will be coming up in the future.

What you can see from the USSB service

With a system of this quality, it was important to ensure that the programming was equally as good. That's why USSB was so careful to hand-pick networks that are not only very popular but also offer a spectrum of programming for an entire family. Look at how they stack up on DSS. We are sure you recognize these networks, and when you see the range of entertainment they offer, you will know why they have become so popular. When all these networks are considered as one complete entertainment package, some surprising facts emerge.

Movies

There are always movies worth watching on USSB. On Showtime and HBO alone, subscribers will see over 80% of the top Hollywood releases in a given time period. Cinemax averages over 190 different movies a month, and more exclusive movies than any other network. Flix runs hundreds of movies you grew up with during the 60s, 70s, and 80s, and the name "The Movie Channel" speaks for itself: It's all movies, all the time. In total, USSB subscribers will be able to choose from over 800 movies every month, giving them the widest choice of movies anywhere.

Kids' shows

Not satisfied with just showing cartoons, Nickelodeon regularly interviews children around the country and has developed a network just for them, with programming that is "kid-tested and kid-approved." A network that produces 49 of the top 50 programs for kids and is a recent winner of daytime Emmy awards, Nickelodeon goes beyond entertainment with innovative programs that educate, encourage, and provide opportunities for kids to volunteer in their communities. In addition to Nickelodeon, USSB offers countless movies and specials geared toward children, such as "The Muppets Take Manhattan" or "Mrs. Piggle-Wiggle" on Showtime, "Old McDonald's Sing-along Farm" on Lifetime, or animated favorites from HBO like "The Legend of White Fang."

Current issues

News and topical issues are well-represented in the USSB lineup. Lifetime, a network dedicated to the interests of women, exposes and deals with topical issues facing women and families today, from breast cancer to child-rearing. MTV, in addition to keeping viewers current on music, creates programs that deal head-on with issues such as violence (through campaigns like "Enough is Enough") and lack of voter turnout (through their "Rock the Vote" campaign.) Finally, for the most current reporting on the issues and the facts that make history around the nation, there's All News Channel. It is a refreshing 24-hour alternative to the other news networks.

Original programming

Much of the programming on the networks selected by USSB is all original material, and cannot be seen anywhere else. Case in point: HBO's highly acclaimed and award-winning original movies, like "The Burning Season," a heroic story of one man who took on the corporate-political machine, or "State of Emergency," nominated for a '94 Cable Ace Award, a movie that explores today's medical care crisis. Showtime, soon to be the world's largest independent producer of theatrical films, delivers original movies directed by Hollywood heavyweights like Tom Cruise, Kathleen Turner, and Tom Hanks. Lifetime contributes original, world premiere movies such as "Against Her Will: The Carrie Buck Story," about the sterilization of mentally deficient women. Comedy Central offers originals like "Mystery Science Theatre 3000," a regular program in which two robots and a space traveler provide continuing comedic commentary on "B" movies. (You have to be there.)

Comedy

At USSB, there is always something funny going on. This is particularly true on Comedy Central, the only 24-hour-a-day comedy channel. Their distinctive view of the world connects with today's tough audiences and keeps them laughing. One show, "Politically Incorrect," with guests like Jerry Seinfeld and Roseanne Barr, has been described as "The McLaughlin Group on acid." Other comedy offerings include HBO's "The Larry Sanders Show" with Garry Shandling, MTV's irreverent "Beavis and Butthead," and many Showtime comedy specials with stars like Brett Butler.

Music

USSB's digital signal and CD-quality sound make it the best vehicle available for music television. And the two channels that created music television are on USSB. MTV, the only TV channel to have a generation named after it, is the world's first day-to-day lifeline between that generation and its music. Since its inception as a music video showcase, MTV has added many original programs of interest to its devoted fans and presents one of the biggest events in music today, the MTV Video Music Awards. VH1 delivers an equally stimulating array of music and entertainment to a slightly older audience of MTV graduates. HBO also treats their viewers to front-row seats at spectacular music events like the "Whitney Houston Concert from South Africa."

Classics

The golden age of television can be relived on USSB. The brightest moments from America's most famous TV shows shine again on Nick at Nite, the home of classic TV and a perfect choice for families that grew up with television. Flix offers a wide variety of classic movies like *The Big Chill* and *Kramer vs. Kramer*. And The Movie Channel features many soon-to-be-classics, like "When Harry Met Sally" and "Peggy Sue Got Married."

Whatever your taste in entertainment, USSB has a multitude of great choices to fit your needs 24 hours a day. And for the first time you can get all these great networks without having to buy a long list of channels that you do not want to watch; and when you first sign on with DSS, you will receive a month's free service with USSB.

Multichannel programming

Not only has USSB selected the best programming, it has made that programming available in a way you have never been able to get before. It's called multichannel programming, a technique so impressive that it has given the networks new names: Multichannel HBO, Multichannel Showtime, Multichannel The Movie Channel, and Multichannel Cinemax.

These networks have created a way to increase the enjoyment and satisfaction of their subscribers by multiplying the number of channels they offer. And USSB has them all! They offer you five channels of HBO, three channels of Showtime, two channels of the Movie Channel, and three channels of Cinemax and Flix. Nowhere can you get more. The effect of this is that you are virtually guaranteed to find a program you want to watch, every time you sit down to watch television.

Besides greater genre choice, these new Multichannel Networks also make it easier to catch special events and see more of the network's original programming. They also give you more chances to catch a favorite movie. In fact, a recent survey found that people overwhelmingly felt that Multichannel Networks gave their family increased programming choices, and many believed they increased the value of their subscription.

By now it should be clear that USSB offers a vast array of programming. The DSS system also has a number of features that keep you from missing what you want to see and, perhaps more importantly, keep others (such as your kids) from watching what you do not want them to see.

Most of this control comes from the DSS feature called locks and limits. With the special remote control, you have the power to set rating limits for your TV. You can select whatever rating limit is right for you and your family, whether it's PG-13 or G. Then you lock the system with a four-digit personal ID code. When you want to watch programs that exceed the rating limit, simply enter your personal ID code and bypass the lock.

The remote also helps you select from the wide variety of programming available. An up-to-date on-screen television guide lists everything that's playing, or about to play. You can also customize the display to list only certain types of programming, such as sports or movies. You can even create a couple "Top 10" favorite channel listings. With the push of a button you can call up a wealth of data on a movie, including its rating, who stars in it, the director's name, and a plot summary. I guess you could say someone is finally putting some real control into the remote.

The USSB and Hubbard services

USSB has the best picture, the best sound, and the best programming ever broadcast. It's only logical that they should work with the best retail and customer service professionals, and USSB does that.

When the groundwork for USSB was being laid, customer service support and satisfaction was a cornerstone. The Hubbards were already selecting and training people even before the Digital Satellite System was up and running.

To begin with, getting a DSS system—or just taking a look at one—is as easy as stepping into your local RCA dealer or satellite TV retailer. These retailers were chosen because of their reputations and expertise in selling and servicing electronics, as well as their being conveniently located across the nation.

Once you get your system home, programming assistance is as close as the phone. A toll-free call hooks you up with the USSB Customer Service Center, where professionals—real, live, actual people—are ready to answer your questions or to get you someone who can, 24 hours a day, 365 days a year. Any question about your bill or about how to get the most out of your programming or your subscription will be answered quickly and completely.

But none of this means anything unless you check out USSB for yourself. Stop by one of our retailers and ask for a demonstration. Once you take the remote in hand and experience the power of digital entertainment, we are sure you will agree this is one of the greater advances in TV entertainment.

The Hubbard family

The Hubbards made sure USSB had the most acclaimed programming being created, programming that not only entertained the whole family, but advanced the legacy of television itself.

■ **8-1** *Mary Pat Ryan, senior vice president of marketing for USSB.*

Then they sought out the most perfect way to broadcast these channels. The answer lay in the new RCA Digital Satellite System (DSS). The way DSS works is simple, if sophisticated. The photo that is Fig. 8-1 is of Ms. Mary Pat Ryan, senior vice president of marketing for USSB.

USSB digital entertainment

In Figs. 8-2, 8-3, 8-4, and 8-5 on the following pages you will find the logos and information for the various entertainment program channels that you will find on USSB.

Unfettered by the restraints of broadcast TV, HBO has blossomed beyond the blockbuster movies they started showing two decades ago. Their world-class original programming now gives the HBO mantlepiece the glint of Emmy and Oscar.

Serving up a treasure trove of cinematic gems, TMC has achieved an optimal mix of blockbusters, sentimental favorites and a gratifying offering of those 'B' flicks we can't explain why we love, but we do.

"Oh, I haven't seen this for such a long time," is most often heard when looking through choices on FLIX. They weave together the best films from the '60s to the '90s with programs built for people who not only love to watch movies, but love to know about them.

■ **8-2** *Information on HBO, Multichannel, and Flix.* USSB

Not content to just celebrate women as they are, Lifetime seeks to better their condition. Their recent original movies and support of causes such as breast cancer prevention and child care are apt examples that TV can indeed be a positive force for change.

Lifetime™
Television for Women

With a philosophy of "kids deserve the best," they quickly became the largest creator of kids' programming in the world. Now the home of 49 of the 50 most popular kids shows, they continue to teach kids that learning is a fun part of life.

If laughter is the best medicine, then this is America's new health plan. From the stars of stand-up to the razor-sharp wit of their political parodies, COMEDY CENTRAL has shown us that we laugh hardest when we can laugh at ourselves.

■ **8-3** *Lifetime, Nickelodeon, and Comedy Central channels.* USSB

Realizing that more than big names and big studios make good movies, SHOWTIME adds to their exclusive blockbuster list by offering releases from smaller and more daring independent producers.

Averaging over 170 different movies a month with the fewest repeats of any premium service, CINEMAX continues to show the winners that others miss. Which makes this network a delightful surprise every time it's tuned in.

The cultural lifeline of today's youth has found that its voice speaks with power and cutting-edge vibrancy in more than music. With its maturing shows and specials, MTV helps steer them through their turbulent times by showing that, yes, we understand.

■ **8-4** *Showtime and Cinemax, plus Music TV.* USSB

Music speaks to all generations. Today's adults feel the same power from the sound as when they were young. In fact, it keeps them young. No other network embraces that feeling through contemporary music like VH1.

TV is an integral part of the American experience. We all have memories from the screen that pleasantly linger with us today. And this is the place where the best and brightest from those golden years have the chance to shine again.

24 hours a day, 5,000 journalists from 150 news bureaus around the globe bring you breaking national and world news, plus the latest on weather, sports, business and entertainment.

■ **8-5** *The VH1, Nick at Nite, and the All News Channel.* USSB

What you can watch with the DSS dish

DirecTV programming

In Fig. 8-6 you will find a list of preliminary DirecTV programming information.

DIRECTV PROGRAMMING

The preliminary DIRECTV programming line-up includes:

CABLE PROGRAMMING SERVICES

THE CARTOON NETWORK	CMT: COUNTRY MUSIC TELEVISION
CNN	CNN INTERNATIONAL
COURTROOM TELEVISION	C-SPAN AND C-SPAN 2
THE DISCOVERY CHANNEL	THE DISNEY CHANNEL
E! ENTERTAINMENT TELEVISION	ENCORE
ENCORE THEMATIC MULTIPLEX	ESPN
THE FAMILY CHANNEL	HEADLINE NEWS
PLAYBOY TV	THE GOLF CHANNEL
THE LEARNING CHANNEL	THE SCI-FI CHANNEL
THE TRAVEL CHANNEL	THE WEATHER CHANNEL
TBS SUPERSTATION	TNN: THE NASHVILLE NETWORK
TURNER CLASSIC MOVIES	TURNER NETWORK TV (TNT)
USA NETWORK	

PAY PER VIEW MOVIES

PARAMOUNT PICTURES	COLUMBIA PICTURES
SONY PICTURES CLASSICS	TRISTAR PICTURES
TURNER MGM FILM LIBRARY	UNIVERSAL PICTURES
WALT DISNEY PICTURES	MIRAMAX PICTURES
HOLLYWOOD PICTURES	TOUCHSTONE PICTURES

■ **8-6** *A DirecTV programming information listing.* DirecTV

DirecTV special interest programming

Newsworld International

The Canadian Broadcasting Company will create a new-24-hour international news service devoted to comprehensive news coverage and hard-hitting current affairs features.

Northstar

This is a new family-oriented entertainment service from Canada that will feature a compelling mix of drama, arts, light entertainment, and journalism.

Physician's Television Network (PTN)

This network provides specialized programming for physicians that includes medical symposia, debates on key medical issues, and features on medical specialties and the legal aspects of practicing medicine. Through this service, physicians can receive continuing medical education credit from the Network for Continuing Medical Education (NCME) program sponsored by Visual Information Systems, Inc. (VIS).

DirecTV will also feature additional special-interest programming, including education, cultural programming, foreign languages, and programming targeted at professional and affinity groups across the United States.

Hughes-Hubbard and DBS

Two companies have invested billions on the DBS venture, and they share these satellites. They are the Hubbard's United States Satellite Broadcasting and DirecTV, a division of the huge GM Hughes family of companies.

Hubbard Broadcasting began with a single Minnesota radio station back in 1923, but today it is poised to work a revolution on the communications world.

The Hubbard Broadcasting arm of United States Satellite Broadcasting (USSB), will be aboard the nation's first direct-broadcast satellite. In April of 1994, it went on line offering at least 20 digital channels of programming. USSB will share the satellite known as DBS-1 with DirecTV, an outgrowth of Hughes Communications Company.

This DBS programming is beamed from the USSB National Broadcast Center in Castle Rock, Co. to DBS satellites stationed 22,300 miles over the earth. Figure 8-7 illustrates this DSS concept. Digital programming is then beamed down from the satellites to the small, 18-inch RCA DSS dish that can be attached to the side or roof of a house. A set-top receiver picks up the programming signals from the dish and puts them onto your TV receiver.

So far, the Hubbard family has sunk vast sums of investment, would you believe tens of millions of dollars, into this project. The contract with Hughes, which gave Hubbard ownership of five of the 16 transponders on the first satellite, is worth at least $100 million, according to USSB President Stanley E. Hubbard II. Fig-

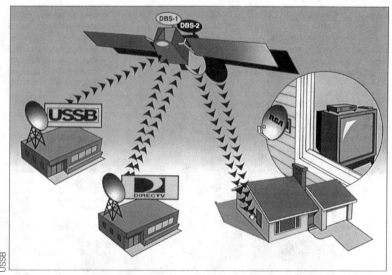

Digital programming will beam down from the satellite to your DSS dish, and then flow to the IRD in your home via copper coaxial cable.

■ **8-7** *The total DBS delivery concept.*

ure 8-8 on the following page shows President Hubbard in the master control room of the USSB National Broadcasting Center in Oakdale, Minn.

Estimating conservatively, Hubbard says USSB needs to attract three million viewers to break even. With some luck, he says, the company could break even at one million. It could be that cable TV will lose viewers once they see the DSS alternative programming.

Although the Hubbards were in the DBS game before Hughes, they now find themselves at a potential disadvantage. DirecTV has 27 of 32 possible transponders at the 101 degrees west longitude orbital slot, 11 on the first satellite and 16 on the second satellite. DirecTV is also part of the Hughes empire—from aircraft to satellite construction—as well as being into TV program syndication. USSB, therefore, may be in danger of being lost in the shuffle of its large DBS neighbor.

The trick is to create a separate market identity, says Hubbard. USSB is staking its name on a package of channels heavily weighted toward premium services. The centerpieces of the lineup, which will include at least 20 channels, are HBO, Showtime, and their multiplexed variations. When the service goes on the air in April, HBO will have five distinct channels and Showtime will have three. USSB will add six movie channels, includ-

166

USSB

■ **8-8** *Stanley E. Hubbard in the USSB control room.*

ing Flix, for a total of 14 premium channels. The remaining services include MTV and Nickelodeon.

In a bow to Hubbard Broadcasting's heritage as a TV and radio owner, USSB will offer one or two free, advertiser-supported channels, along the lines of traditional TV stations or satellite-delivered superstations such as Turner's WTBS Atlanta. USSB will produce its own channels rather than picking up WTBS or another station.

USSB viewers will see a lot of programs that will be produced and developed by local TV stations, as well as by major Hollywood producers who want to have a window into this market and want to start out and grow with this new business venture.

Despite the effort to differentiate USSB from DirecTV, the fortunes of the two companies are intricately linked. Because the Thomson receiving system works for both transmissions, the companies have a vested interest in seeing that as many households as possible have receiving equipment. DirecTV, USSB, and Thomson will contribute to a joint marketing fund, to be run by Thomson, to sell the system. Dealers will be able to sign up customers at the point of sale. USSB's program package is designed to be different from DirecTV's and yet complement the competing DirecTV lineup. Many customers will subscribe to all or part of both services; that way, the services split the cost of launching such a service and support each other in developing the market, despite being competitors. The arrangement is comparable to regular TV stations that have a common tower. One company will own the tower and lease out tower space to other broadcasters.

USSB has been involved in DBS from the first minute in 1981 that the FCC announced it would offer licenses. Stanley S. says he saw the potential immediately and wanted to get in on the ground floor. He saw it as a way to compete with cable, which was just beginning its meteoric rise.

"I have never accepted for one second, nor do I today," he says, "that having to be connected to a wire is a step forward. To me, having to be connected to a wire is a step backward." Although broadcasters were leery of Hubbard's plan, they are now showing signs of enthusiasm for DBS. Not only is it potential competition to cable, it is also potentially their gateway to multichannel programming through services such as USSB's proposed advertiser-supported superstation services.

Figure 8-9 on the following page is a photo of the USSB digital pioneers (left to right) Stanley E. Hubbard, president and chief operating officer; Robert W. Hubbard, executive vice president; and Stanley S. Hubbard, chairman and chief executive officer, gathered around an RCA retail kiosk display. USSB was founded by Hubbard Broadcasting, Inc. in 1981 to pursue the corporate vision of creating the first Direct Broadcast Satellite national broadcast service in the U.S. The DSS kiosk symbolizes the culmination of USSB's 14-year-long quest to deliver the highest-quality national broadcast service, and make it available to every American household.

■ **8-9** *The USSB pioneers around the RCS DSS kiosk display.*

Stanley E. Hubbard (standing on raised platform in Fig. 8-10), president and chief operating officer, and Robert W. Hubbard (on the stairs), executive vice president, stand alongside one of the two state-of-the-art digital broadcasting transmitter dishes installed at the USSB National Broadcast Center in Oakdale, Minn. The transmitter dishes are almost 30 feet (9 meters) in diameter. High-powered digital television signals are transmitted via the two dishes to the DSS broadcast satellite 22,300 miles above the earth. The 18-inch RCA DSS satellite receiver picks up the return signal for home reception of USSB's 20-channel array of premier-quality digital TV.

DSS questions and answers

The following questions and answers will help the TV viewer that is contemplating purchasing a DSS satellite dish and receiver.

USSB

■ **8-10** *Stanley and Robert Hubbard with the digital USSB transmitter uplink dishes at the Oakdale, Minn. center.*

What is DSS?

DSS stands for Digital Satellite System, a new, state-of-the-art digital television service launched in North America by USSB, a subsidiary of Hubbard Broadcasting; DirecTV, a unit of Hughes

Electronics Corp; and RCA/Thomson Consumer Electronics. DSS allows households to receive approximately 175 channels of digital television programming directly from two high-power satellites. Whether in remote rural areas or in a major city, homes receive the signal on an 18-inch satellite dish. As of October, 1994, DSS was available to consumers nationwide at over 10,000 retail locations.

What are the benefits of DSS?

Choice, quality, and value, in a nutshell. DSS is a true national broadcasting system, available to everyone in the continental United States on an equal basis. It is a spectacular choice in television entertainment, providing convenience for people because it is so incredibly easy to use. Viewers use a remote control to select the programming they want from an on-screen menu. It's that simple.

With approximately 175 channels, all broadcast in digital video and sound, DSS enhances the viewing experience much as the compact disc improved home audio systems. You will find it crystal clear, rivaling the sound clarity of CDs, while the pictures are as flawless as those from a laser disc.

DSS offers more choices than any other type of commercial broadcast system. From USSB alone, viewers can receive more quality entertainment choices than were ever available to them before. With five multichannels of HBO, three multichannels of Showtime and Flix, two multichannels of the Movie Channel, three multichannels of Cinemax, Lifetime, Comedy Central, Nickelodeon/Nick at Nite, MTV, and the All News Channel, USSB is a home entertainment lover's dream come true.

There is more yet. DSS also offers programming services from two distinctive program providers, USSB and DirecTV. DirecTV offers a complementary programming lineup that includes 40–50 pay-per-view channels of Hollywood movies, extensive sports programming, and a wide array of additional special interest programming.

The effect of this is that consumers are virtually guaranteed to find a program they want to watch, every time, thanks to the availability of multichannel premium channels and pay-per-view programming. The DSS on-screen viewing guide provides continually updated program listings and descriptions, and allows viewers to scan ahead to see program schedules as much as 24 hours in advance. DSS also features an innovative parental control system that allows DSS owners to establish limits and control pay-per-view spending.

USSB has a multitude of great choices to fit consumer needs 24 hours a day. And for the first time, people can get all these great networks without having to buy a long list of channels they do not want.

What does it cost?

A complete DSS hardware unit can be purchased at a starting basic unit price of $699. Monthly programming packages range from $5.95 to $34.95, depending on what the viewer wants to watch. In many cases, a consumer may want to purchase packages from both USSB and DirecTV.

The cost breakdown for hardware system options is as follows:

Basic system ($699)

This system allows for independent service hookup to one television.

Deluxe system ($899)

This system allows for multiple television hookups. It also includes a universal remote control.

Second satellite receiver ($649)

A second receiver provides independent signal for a second television set.

What programming is available on DSS?

USSB currently offers viewers 20 channels, including multichannel versions of the most popular premium services such as five multichannels of HBO, three multichannels of Showtime and Flix, three multichannels of Cinemax, two multichannels of The Movie Channel and a hand-picked selection of quality "basic" channels, including Lifetime, Comedy Central, Nickelodeon/Nick at Nite, MTV, VH1 and the All News Channel.

DirecTV offers approximately 150 selections of popular networks, hit pay-per-view movies, professional and collegiate sports, and special interest programming. The $21.95 DirecTV Direct Choice package includes channels such as CNN, ESPN, The Disney Channel and TBS, among others. A variety of other DirecTV packages ranging from $5.95 to $29.95 are also available.

☐ 14 premium movie channels
☐ 40–50 major subscription channels

☐ 30–40 sports channels

☐ 50 or more pay-per-view (PPV) movie channels

How is the DSS hardware installed?

The DSS hardware is easy to install. The dish can be installed in almost any outdoor location, as long as it has a direct line of sight to the south. Professional installation is readily available, or handy do-it-yourselfers can install and position their own dishes. All programming is received from one position in the sky, and the dish is equipped with a circular-polarized transmission system. Thus, the dish remains in one place, minimizing maintenance and eliminating the need for satellite tuning and repositioning.

The satellite receiver displays the proper dish elevation for any area when a zip code is entered. Installers level the mount, then use the elevation scale built into the mount to set the dish angle above the horizon. The system has a built-in audiovisual signal meter, so installers can pan the horizon watching and listening to the strength of the signal tone to position the dish properly.

Consumers can install their own dish with a "do-it-yourself" installation kit available for $69.95 at any authorized DSS dealer. Outdoor installation requires drilling a hole in the house's facade and running cable through the walls to connect the service. The price for basic professional installation is approximately $150 to $200.

How does the on-screen menu work?

After subscribing to USSB, DirecTV, or both services together, the viewer uses a remote control to pull up a menu on the television screen and select a program. The menu has categories of programs (sports, movies, etc.) with submenus (e.g. comedy, adventure, romance, etc.) The menu also provides a program guide so viewers can scan what's on at any time. DSS channels are numbered from 100–999; the menu conveniently allows consumers to select the programs they want.

What about local TV broadcasts?

Consumers will still be able to receive local stations in the same way they receive them now. USSB and DirecTV are national programming distribution services, meaning local channels are not included as part of the program lineups. To provide consumers with a complete viewing experience, DSS has been specifically designed to easily integrate local broadcast channels into the system. In markets where broadcast or cable systems are in place, consumers can maintain either a "lifeline" cable service (usually between $7 and $10) or connect a broadcast antenna to the DSS digital receiver to receive local and network broadcasts. A simple "A-B" switch, built into the DSS remote control, enables consumers to easily and instantly switch between DSS and local stations.

However, DirecTV offers one network affiliate from each of the five major networks (ABC, NBC, CBS, FOX, and PBS) to viewers in areas unserved by broadcast or cable television. As part of the Satellite Home Viewing Act of 1988, the Federal Communications Commission (FCC) prohibits the delivery of network affiliates to areas with their local network affiliates.

How are customers billed?

Customers are billed individually for their subscriptions by USSB and DirecTV. Pay-per-view movies are billed automatically through the satellite receiver. Viewers subscribing to both services receive two bills each month: one from each company.

Why does the system require an access card?

The credit card-sized access card that comes with each system acts as a "license plate," providing security and encryption information, allowing customers to control the use of DSS, and enabling programming providers to capture billing information.

Where does DSS broadcasting originate from?

USSB and DirecTV each have their own broadcast centers. The USSB National Broadcasting Center is a 20,000 square-foot uplink facility located in the Twin Cities area of Minnesota. USSB uses two nine-meter Ka-band transmit dishes and multiple C and Ku band downlinks. DirecTV broadcasts from its national uplink center in Castle Rock, Colo. USSB's National Broadcast Center is linked to Hubbard Broadcasting's Twin Cities headquarters, which houses All News Channel, Conus Communications' satellite news gathering and television production center, and KSTP-TV's television studios and extensive production resources. Satellite links provide access to television production facilities at Hubbard Broadcasting's F&F Productions, Diamond P Sports and WTGO-TV in St. Petersburg, Fla.; Conus in Washington, D.C.; KOB-TV and Pro-IV Productions in Albuquerque; and WDIO-TV in Duluth, Minn.

How does DSS compare with C-band satellites?

C-band and DSS appeal to two very different audiences. With its larger dish, C-band appeals to enthusiasts and video hobbyists desiring unlimited access to all forms of satellite television transmissions. DSS is designed as a mass-market television delivery system for the average American television viewer. With DSS, viewers receive the benefits of a multichannel broadcast system with lower equipment costs, smaller size, and less maintenance than a conventional satellite TV system.

How does DSS compare to cable?

DSS is the next dramatic step in the evolution of television—a tremendous entertainment value. No other system offers as many quality programming options (up to 30 channels from USSB and 150 channels from DirecTV) at any price. The programming from both companies is complementary, giving consumers an astonishing range of digital programs to choose from. For example, USSB delivers multichannel versions of the most popular premium subscription services. Consumers can choose from among five channels of HBO at once, each of which airs a different entertainment program at the same time. Both USSB and DirecTV offer special interest programming, in addition to DirecTV's Direct Ticket, the first commercially viable pay-per-view delivery system.

While DSS programming rates are comparable to cable, DSS gives viewers the unique ability to control their costs. Unlike cable, which is sold in rigid packages, DSS programming is structured to be more flexible so that customers pay for the programming they want and have more control over the amount they spend for TV viewing.

Is DSS affected by inclement weather?

DSS is designed to achieve a minimum 99.7 percent reliability rate. In most areas, reception reliability is closer to 99.9 percent—a rate comparable to that of conventional on-air television network standards.

Normal rain will have no effect on a properly installed DSS receiver system. DSS reception is affected only by the most severe weather, and then only for very brief periods of time. Service will quickly return to its normal high level of quality by itself, without the need for a service call by a technician.

Is DSS an interim technology to be replaced by the information superhighway?

No. DSS is fully digital and "forward compatible" so that consumers can take advantage of emerging video technologies, such as interactive services, 16 × 9 wide-screen TV transmissions, and HDTV broadcasts. The deluxe DSS model has a data port capable of delivering many of the features of the highly-touted "information superhighway."

However, the crucial difference between DSS and the superhighway is the simple fact that DSS is here now. It is commercially available, unlike superhighway multimedia technologies that have been more speculative than real.

According to an estimate published in the *Wall Street Journal*, it would cost, conservatively, $100 billion to deliver a fiber-optic highway to American homes. DSS, using satellite technology to also deliver more information and entertainment, is now in place at a total cost of approximately $1 billion.

Where can consumers and dealers get product information?

For more information, call the following 800 numbers.

USSB consumer hotline	800-BETTERTV
USSB dealer hotline	800-898-USSB
RCA DSS information	800-898-4DSS
DirecTV consumer information	800-DirecTV
DirecTV dealer information	800-323-1994

In Fig. 8-11 on the following page you will find the DSS programming summary. In Fig. 8-12 on the following page is the USSB subscription programming information. You will find the complete DirecTV program information shown in Figs. 8-13 through 8-20 on the following pages, along with the cost for the various packages.

DirecTV customer service

DirecTV offers customers first-rate service and support with well-trained customer service personnel. Their 24-hour customer service center is accessible 24 hours a day, seven days a week.

☐ It directs potential customers to retail outlets.

☐ It provides customers with information on programming selections, billing and sales.

☐ It allows customers to select their own billing method (credit card or monthly statement).

USSB℠ Digital Satellite System (DSS®) Programming Summary
The Best In Digital Entertainment Programming For The Entire Family

Home Box Office	5 great HBO channels of first class, premium Hollywood films, original shows, comedy, family series, concerts, boxing and documentaries.
SHOWTIME and FLIX	3 unique SHOWTIME channels featuring award-winning, critically acclaimed box-office hits, original movies, documentaries, boxing, comedy and family specials, while FLIX features all your favorite movies from the '60s, '70s and '80s, uncut and commercial-free.
The Movie Channel	2 different channels of non-stop, commercial-free movies, including classic romance, adventure, mystery, drama and suspense hits.
Cinemax	3 distinct channels specializing in a variety of movies categorized for easy selection, seven days a week.
MTV	The cutting edge in music and pop culture.
VH1	The channel where active adults stay connected to the music they love.
Nickelodeon	Kid-tested, kid-approved, the #1 network for kids.
Nick at Nite	Nick at Nite brings families together every night with the best of classic TV.
Comedy Central	The only all-comedy channel. It's the one place to turn to 24 hours a day for smart, surprising, original comedy.
Lifetime	Lifetime offers women entertainment, information and inspiration throughout the day.
All News Channel	Up to the minute national and international news, weather, sports, business and entertainment news.

USSB/NR/112-94/3-9-95

■ **8-11** *A DSS programming summary list.* USSB

An added level of customer-first services are offered by DirecTV, including a printed program listing guide offered on a monthly subscription basis, and on-air customer information. DirecTV simply offers TV viewers more, as compared with other forms of video entertainment. DirecTV offers more:

☐ Quality: The all-digital format provides superior reception, including the capability for laserdisc-quality picture and CD-quality sound.

☐ Variety: There is a broad selection of programming choices, such as pay-per-view movies, cable channel favorites, sports

USSB℠ SUBSCRIPTION PROGRAMMING INFORMATION

USSB (United States Satellite Broadcasting) programming includes hand-picked networks that are not only extremely popular, but also offer a spectrum of programming for an entire family. For the first time, subscribers can get all these great networks without having to buy a long list of channels they don't want. To help with this decision, all DSS• purchasers are invited to enjoy one free month of USSB's top programming package, USSB Entertainment Plus℠. For more information on USSB digital entertainment programming, call 1-800-BETTER-TV.

USSB Entertainment Plus

$34.95 per month:
- HBO, HBO2, HBO3, HBO West, HBO2 West
- SHOWTIME, SHOWTIME2, SHOWTIME West, FLIX
- The Movie Channel, The Movie Channel West
- Cinemax, Cinemax2, Cinemax West
- Lifetime, Nickelodeon/Nick at Nite, MTV, VH1, Comedy Central, All News Channel

HBO Plus

$24.95 per month:
- HBO, HBO2, HBO3, HBO West, HBO2 West
- Cinemax, Cinemax2, Cinemax West
- Lifetime, Nickelodeon/Nick at Nite, MTV, VH1, Comedy Central, All News Channel

SHOWTIME Plus

$24.95 per month:
- SHOWTIME, SHOWTIME2, SHOWTIME West, FLIX
- The Movie Channel, The Movie Channel West
- Lifetime, Nickelodeon/Nick at Nite, MTV, VH1, Comedy Central, All News Channel

Select One Plus

$17.95 per month:
- Choose any of the premium multichannel services offered by USSB — Multichannel HBO; Multichannel SHOWTIME and FLIX; Multichannel The Movie Channel; or Multichannel Cinemax; and add the USSB Basics package.

SHOWTIME Package

$10.95 per month:
- SHOWTIME, SHOWTIME2, SHOWTIME West, FLIX

HBO Package

$10.95 per month:
- HBO, HBO2, HBO3, HBO West, HBO2 West

USSB Basics℠

$7.95 per month:
- Lifetime, Nickelodeon/Nick at Nite, MTV, VH1, Comedy Central, All News Channel

USSB/NR/111-94/4-12-95

■ **8-12** *USSB subscription programming information.* USSB

and pay-per-view special events, commercial-free audio channels, and free preview channels.

☐ Personal choice: Customers can add their favorite cable networks to a variety of programming package selections, creating their own customized entertainment.

☐ Convenience: Popular pay-per-view movies are available as often as every 30 minutes, enabling customers to watch

Direct Ticket™ Pay Per View. Features up to 50 channels of hit movies, sports and special events. Hit films are available as often as every 30 minutes and can be instantly ordered using the on-screen program guide and remote control. Programming is provided by major Hollywood and independent motion picture studios that include:

Turner MGM Film Library

Sony Pictures Classics

Major event promoters

■ **8-13** *A portion of the DirecTV programming listing.* DirecTV

what they want, when they want—without leaving their homes.

☐ Value: DirecTV delivers more of the programs customers want and at a price that is competitive or better than other television entertainment systems.

The USSB company

A co-owner of the DBS-1 satellite, USSB shares the entire DSS digital satellite system. It is managed by Hubbard Broadcasting. Some facts about Hubbard:

☐ They have over 70 years of broadcasting experience.

☐ They own TV and radio stations in Minnesota, Florida, and New Mexico.

☐ They manage Conus Communications, the first and largest satellite news-gathering service in the world.

A&E. The leader in quality entertainment featuring the best in comedy, drama, documentaries and performing arts. A&E leads the industry in providing programming with educational merit and has been awarded more CableAce Awards than any other basic cable network.

America's Talking. Talk is what this interactive, 24-hour service is all about, featuring call-in shows, coast-to-coast roving reports, soap operas and celebrity interviews.

E! Entertainment Television. The only network devoted exclusively to the world of entertainment. Among E!'s compelling features are celebrity interviews, previews of the latest movie releases, and original programs such as *Talk Soup*.

PrimeTime 24. Provides network television service to areas in the United States that are unserved by local affiliates or cable. Its five channels are WABC (ABC-New York), WXIA (NBC-Atlanta), WRAL (CBS-Raleigh), WFLD (FOX-Chicago) and KRMA (PBS-Denver).

Turner Network Television (TNT). When it comes to big-time entertainment, nobody plays movies, makes originals, knows 'toons or covers sports like TNT. Programming features the greatest movies Hollywood ever made, star-studded original productions, and classic kids' shows.

Superstation TBS. Features an outstanding array of family-oriented programming. See exclusive specials, compelling documentaries and movies, plus professional sports.

USA Network. USA features syndicated dramas like *MacGyver*, *Murder She Wrote* and *Counterstrike*. These shows, in addition to comedy programs, variety specials and exclusive sports coverage, make USA one of the most popular cable networks in prime time.

■ **8-14** *DirecTV general entertainment.* DirecTV

Bloomberg *NEW*
DIRECT

Bloomberg Direct. Provides continuous 24-hour coverage of worldwide business and financial news. Forty news bureaus, all major commodities/debt/equity exchanges and the U.S. Chamber of Commerce contribute original breaking news stories, statistical data and training seminars.

CNN.

CNN. CNN provides viewers the fastest, most complete 24-hour coverage of breaking news. It offers a variety of programs ranging from business to sports to entertainment, as well as topical interviews and the highly-regarded *Larry King Live* program.

CNN *NEW*
INTERNATIONAL.

CNN International. 24-hour international news service features special events, sports and weather, plus the ability to "go live" to the scene of late-breaking news stories from every corner of the world. Available in 1995.

CNBC

CNBC. Financial news highlights dominate the day, while nights include features and discussions of contemporary business issues.

C-SPAN   C-SPAN2

C-SPAN and C-SPAN2. Unique news and information programming, including 24-hour coverage of important political events from Washington, DC, and around the nation. C-SPAN offers debate from the floor of the U.S. House, live and in its entirety. C-SPAN2 covers the U.S. Senate.

COURT TV
COURTROOM TELEVISION NETWORK

Court TV. The only 24-hour network dedicated to live and taped coverage of courtroom trials from around the U.S. The coverage is supplemented by programs that focus on courts and legal issues in America and around the world. As many as three live trials are covered daily.

Headline
NEWS.

Headline News. Every 30 minutes, Turner Broadcasting's 24-hour news service delivers an updated, concise report on the day's top stories, business, sports and entertainment news for the on-the-go viewer.

■ **8-15** *DirecTV news and information.* DirecTV

Newsworld International. The renowned Canadian Broadcasting Corp. supplies 24-hours of international coverage devoted to hard-hitting news and comprehensive current affairs features.

The Travel Channel. Business and leisure travelers find this 24-hour network a valuable source of great ideas for all kinds of travel. Original programs feature tourism experts, authors, newsmakers and celebrities.

The Weather Channel. Weather watchers can find 24-hour reports on regional and national weather conditions. Special weather-related features and reports on unusual weather phenomenon.

■ **8-15** *Continued.*

☐ They are co-owners with VIACOM of the All News Channel, a 24-hour satellite-delivered news channel.

☐ They own Diamond P sports and F&F Productions.

☐ They have been pioneering DBS (Direct Broadcast Satellite) since 1981, and are the longest-standing licensed DBS operators.

☐ They formed a strategic alliance with Hughes/DirecTV to share satellites and the entire DSS digital satellite system.

☐ They formed a strategic alliance with Thomson Consumer Electronics to jointly distribute hardware and USSB programming.

USSB programming information

USSB has the power of Multiplex, which enables them to offer multiple channels of a premium service, much like a cineplex movie theater. USSB has 14 premium movie channels, including five of HBO, three of Showtime, three of Cinemax, two of The Movie Channel, and one of Flix.

☐ Multiplex = Greater proven customer satisfaction

☐ Multiplex = Increased value for the customer

☐ Multiplex = More channels for the same price

☐ Multiplex means more choice, variety, quality, and value.

ESPN. America's Number One sports network delivers all sports, all the time, plus diverse sports-related news and information. Features Sunday Night NFL Football, Major League Baseball and NCAA Basketball.

 NEW

ESPN2. An exciting, fast-paced mix of sports events, news, information, sports entertainment and nightly themed blocks for the younger sports fan.

 NEW

THE GOLF CHANNEL

The Golf Channel. 24-hour live and tape-delay coverage of world-class international tournaments from Europe, Africa and Australia. Available in 1995.

Turner Network Television (TNT) sports. Great programming includes sports and weekly series, plus the ever-popular NBA on TNT and Sunday Night NFL Football. Not available in all areas.

Superstation TBS sports. One of America's most popular basic entertainment networks includes a wide variety of professional sports and sports specials.

THE NASHVILLE NETWORK

TNN: The Nashville Network sports. Wide-ranging programming includes extensive coverage of the NASCAR racing circuit.

USA Network sports. Exclusive programming for die-hard sports fans includes championship tennis and golf events. USA is also the official cable network of the World Wrestling Federation.

 NEW

ESPN/ABC Sports College Football Pay Per View. Exclusive coverage of the country's top-ranked college football games not usually available locally on broadcast TV. Features teams from all the top conferences. Available on a seasonal or individual weekend basis.

■ **8-16** *DirecTV sports.* DirecTV

NFL SUNDAY TICKET™. Professional football games from the NFL every Sunday during season! In 1994, a package of up to 12 regular season games every Sunday for five weeks — over 50 games in all — beginning November 27. Approximately 200 regular season games will be available for the 1995 season.

■ **8-16** *Continued.*

Cartoon Network. The world's first and only 24-hour network offering the best in animation programming. For cartoon lovers young and old, this network offers 8,500 animated programs from the Hanna-Barbera libraries including *Bugs Bunny*, *The Flintstones* and many more.

The Discovery Channel. Explore your world with powerful and insightful news and information documentaries from the worlds of science, nature, medicine and outdoor adventure.

The Disney Channel

The Disney Channel. Quality entertainment for the whole family, featuring animated Disney classics, original series, entertainment specials and movies.

The Family Channel. Highlights positive values and offers a broad mix of original programs and classic favorites for the entire family. Program scheduling includes a Saturday block of westerns, daily children's shows, dramas, sitcoms and movies.

T L C THE LEARNING CHANNEL'

The Learning Channel. Entertaining and informative programming 24-hours a day, including six commercial-free hours of daily programming for pre-schoolers.

■ **8-17** *DirecTV family and children viewing.* DirecTV

183

Sci-Fi Channel. Features the best of science fiction, science fact, fantasy and horror. The Sci-Fi programming mix includes classic and current popular series, original movies and series, animation, documentaries, plus feature films packaged into festivals and theme weeks.

NEW

TRIO. A family-oriented entertainment service from Canada that features drama, arts and journalism.

■ **8-17** *Continued.*

NFL SUNDAY TICKET™

▶ $139.95 for 1995 regular season package*

- Approximately 200 NFL regular season pro football games‡

- Available for purchase now

- Games usually not available on local cable or broadcast TV channels

ESPN/ABC Sports College Football

▶ $9.95 per weekend

- Three to five out-of-market college football games every Saturday in season

- Top conferences include Big Ten, Pac 10, Big 8, SEC, SWC, ACC, WAC and Big East

NBA LEAGUE PASS℠

▶ $149.00 for '94/'95 regular season package

- More than 400 out-of-market pro basketball games‡

- Games usually not available on local cable or broadcast TV channels

- '94/'95 season now underway

■ **8-18** *Customized packaging.* DirecTV

DIRECTV Sports Choice™

▶ $79.95 per year/$7.95 per month‡‡

- Package of eight out-of-market regional sports networks
- Approximately 100 college football games, 300 college basketball games and 150 college baseball games per season
- Coverage also includes boxing, pro tennis and golf, soccer, volleyball, horse racing, wrestling and much more

A La Carte Networks

- Playboy TV – $4.99 per night, or $9.95 per month
- TV ASIA – $5.95 per month
- PrimeTime 24 – $3.95 per month for ABC, NBC, CBS, FOX and PBS, or 99¢ per month for individual affiliates (available only in areas not served by the local broadcast network affiliates)
- The Golf Channel — $6.95 per month (available January 1995)
- The Disney Channel, ENCORE Multiplex, STARZ!, Music Choice and local regional sports networks are also available on a monthly basis

■ **8-18** *Continued.*

▶ $29.95 per month:

- A&E
- America's Talking
- Bloomberg Information Television
- Cartoon Network
- CNBC
- CNN
- CNN International
- Country Music Television
- Court TV
- C-SPAN
- C-SPAN2
- E! Entertainment Television

■ **8-19** *DirecTV's total choice listing.* DirecTV

185

- ENCORE Multiplex (themed movies)
    ENCORE: '60s, '70s and '80s
    ENCORE2: Love Stories
    ENCORE3: Westerns
    ENCORE4: Mystery
    ENCORE5: Action
    ENCORE6: True Stories and Drama
    ENCORE7: WAM!
- ESPN
- ESPN2
- Headline News
- MuchMusic
- Music Choice (digital audio)

Blues	Hit List
Classical Favorites	Jazz
Classics in Concert	Jazz Plus
Classic Rock	Love Songs
Contemporary Christian	Metal
Contemporary Country	Modern Rock
Contemporary Jazz	New Age
Country Gold	Progressive Country
Dance	Reggae
Eclectic Rock	Rock Plus
Easy Listening	Singers and Standards
For Kids Only	Soft Rock
Gospel	Solid Gold Oldies
Hip Hop	Urban Beat

- Newsworld International
- Regional sports network
- Sci-Fi Channel
- STARZ! — free for 30 days†
- Superstation TBS
- The Discovery Channel
- The Disney Channel – East
- The Disney Channel – West
- The Family Channel
- The Learning Channel
- The Nashville Network
- The Travel Channel
- The Weather Channel
- TNT
- Turner Classic Movies
- TRIO
- USA Network

■ **8-19** *Continued.*

Direct Choice™

▶ $21.95 per month:

- Bloomberg Information Television
- Cartoon Network
- CNBC
- CNN
- Country Music Television
- Court TV
- C-SPAN
- C-SPAN2
- E! Entertainment Television
- ESPN
- ESPN2
- Headline News
- MuchMusic
- Superstation TBS
- The Discovery Channel
- The Disney Channel – East
- The Learning Channel
- The Nashville Network
- The Weather Channel
- TNT
- USA Network

DirecTV Limited

▶ $5.95 per month:

- Bloomberg Information Television
- Access to all pay per view programs and a la carte sports and entertainment offerings

■ **8-20** *Listings of the DirecTV's Direct Choice and DirecTV Limited subscription offerings.* DirecTV

USSB programming essentials

USSB programming has strong appeal to viewers. It includes programming for the entire family, including "Essential" package channels that appeal to each family member, in all age groups. These channels are:

- [ ] Lifetime
- [ ] Nickelodeon/Nick at Nite
- [ ] MTV
- [ ] VH-1
- [ ] Comedy Central
- [ ] All News Channel

Most viewers like these channels as a stand-alone "Essential" package and as the perfect complement to USSB's 14 premium movie channels.

USSB Premium Plus

USSB's 14 premium movie channels offer 14 distinctly different choices at any given time, plus the USSB "Essentials" package. The cost is $34.95 per month, and there is a complimentary one-month subscription with every Digital Satellite System purchased.

USSB customer service center

The USSB customer service center can be reached by dialing these toll-free numbers:

- [ ] 1-800-204-USSB (8772) consumer line
- [ ] 1-800-898-USSB (8772) dealer line

The lines are open 24 hours a day, seven days a week, and will connect the caller to the national hub for:

- [ ] The USSB programming activation center.
- [ ] The USSB customer billing center.
- [ ] The USSB retailer sales commission tracking center.

USSB customer service features top-notch representatives, supported by years of telemarketing expertise. USSB customer service reps will receive ongoing training to ensure that you and your customers get friendly, reliable, high-quality service. The USSB national broadcast center is located in Oakdale, Minnesota.

USSB dealer support

A complimentary one-month subscription to USSB Premium Plus is included with every DSS system purchased, in order to help close the sales. Dealer referrals are available for interested potential customers, and customer communications will be available to the customer from USSB, containing their dealer's name, phone number, and other details. Consumer advertising (in TV, national magazines, and newspapers) is designed to increase awareness, build demand, and get customers interested in the DSS system.

USSB news release

Let's now review the digital dish system and the USSB entertainment experience.

When you purchase the new RCA brand DSS system, you will experience one free month of USSB Entertainment Plus, which is a package consisting of hand-picked networks that are not only extremely popular, but also offer a spectrum of programming for an entire family; and it's absolutely free, and without any obligation.

USSB Entertainment Plus—a $34.95 value—delivers into the home five Multichannels of HBO, three Multichannels of Showtime and Flix, two Multichannels of the Movie channel, and three Multichannels of Cinemax in addition to other family favorites such as Lifetime, Comedy Central, Nickelodeon/Nick at Nite, MTV, VH1, and the All News Channel.

This is a great opportunity for customers to experience the multitude of great choices USSB offers. For the first time, viewers can get all these great networks without having to buy a long list of channels that they do not want to receive.

The free USSB Entertainment Plus offer is available to anyone who purchases the DSS hardware, even if it's already installed; no matter whether it was purchased from an RCA dealer, a home satellite retailer, or the local rural utility cooperative. For the latest details, call toll-free at 1-800-BETTER-TV.

USSB is committed to providing quality for every member of the family, and offers the nation's finest television networks. USSB's programming packages, ranging from $7.95 per month to $34.95 per month, are available through the DSS system, along with complementary programming offered by DirecTV, a subsidiary of Hughes Electronics Corporation.

DSS viewer satisfaction

The over one million consumers who have purchased DSS units based on promises of an unparalleled television viewing experience are reporting overwhelming satisfaction with their entertainment investment.

According to research commissioned by USSB, 97 percent of those interviewed say DSS picture and sound quality have met or exceeded their expectations. Also, the research shows that 84 percent say the programming available on DSS has met or exceeded their expectations.

Another poll of DSS owners indicates that the expectations of quality and variety are the top reasons consumers are buying the 18-inch satellite dishes. Nearly nine of ten (87 percent) of DSS owners rate the "variety of channels" as the top reasons for purchasing the units, while 85 percent say the high-quality picture DSS offers is "very important" or "extremely important." 93 percent say that TV is their primary source of entertainment, and find DSS the best TV viewing available.

Internet access with DirecTV

Sometime in early 1996, GE Hughes DirecTV plans to use the DSS system to offer broadband interactive data services that will have access to the Internet on-ramp.

If DirecTV's plans hold to their schedule, it will be the first company to have economical interactive broadband data services to all U.S. consumers, national in scope.

This will be bad news for cable television companies that have been burning the midnight oil trying to work out the technical bugs that would allow them to offer interactive computer services over the existing analog cable networks already in place. If all goes according to plan, this should put DirecTV ahead of the cable people in the same manner as they are doing with digital video TV into the home. The cable companies are only doing market testing, while DirecTV is proceeding onto the Information Superhighway as a national service.

Types of service

The plan is for DirecTV to offer two types of services. The first one is designed to work with a television receiver while you are viewing video programs. Essentially, this mode will offer the option to overlay text and graphics on incoming television programs. As an example, if you are watching a baseball game, you will be able to bring up on the screen a summary bio of a particular player.

This data will come in almost instantaneously. A tap of the remote, then the entry of a code, and the data flows across the screen.

Another up-and-coming feature will be news services for personal computer users. You will be able to connect the DSS satellite receiver dish directly from the built-in port to your computer. With the built-in telephone modem, you will have the ability to access information just like you do from any online service today. However, the advantage is you will be able to download from the satellite, which will be tremendously faster; as an example, a 1 GB video file could be downloaded in only a few seconds.

The first DSS receivers sold have had the data port option that allows a high-speed computer connection. However, DirecTV plans to offer a new DSS receiver on a plug-in card that would allow the personal computer to be directly connected to the satellite dish.

At this time, DirecTV is meeting with the hardware and software provider folks to work out how the system will look. It will be a full multimedia service that will be as fast or faster than any of the proposed cable modems. The company that will first offer on-line information through DirecTV has not been revealed at this time. Probably, the online service that offers a gateway to the Internet will occupy the bandwidth of about two conventional television channels on the DirecTV system.

At this time DirecTV will not release any technical details of this new service, but there will probably be some announcements in 1996. However, the services being offered by Hughes Network Systems do provide some insight as to how the delivery data system might work. Just stay tuned!

The information, photos, and illustrations in this chapter are courtesy of DirecTV and USSB.

Basic DSS technical overview and troubleshooting

9

THIS CHAPTER BEGINS WITH THE DSS SATELLITE BASICS and operation, then a technical overview of the total DSS system is fully explained (this includes technical information and instructions for operating the DSS digital receiver). Then we will look at some system troubleshooting of the DSS system, learn how to locate faulty phone lines or connections, and troubleshoot poor coax connections, DSS receiver problems, or problems with the satellite dish. The chapter concludes with a DSS system installation certification test that you may wish to take.

Satellite communication basics

All communications services, from ship-to-shore radio communications, radio, and television to communications satellites, are assigned unique bands of frequencies within the electromagnetic spectrum in which they are to operate.

To receive signals from the earth successfully and relay them back again, satellites use very high frequency radio waves operating in the microwave frequency band—either the C-band or Ku-band. C-band satellites generally transmit in the frequency band of 3.7 to 4.2 GHz, in what is known as the Fixed Satellite Service band or FSS. However, these are the same frequencies occupied by ground-based point-to-point communications, making C-band satellite reception more prone to interference.

Ku-band satellites may be classified into two groups, those groups being low- and medium-power Ku-band satellites. They transmit signals in the 11.7 to 12.2 GHz FSS band; and the new high-powered

Ku-band satellites transmit in the 12.2 GHz to 12.7 GHz Direct Broadcast Satellite service (DBS) band.

Unlike C-band satellites, these newer Ku-band DBS satellites have exclusive rights to the frequencies they occupy, and therefore have no microwave interference problems. The RCA DSS system will receive programming from high-power Ku-band satellites operating in the DBS band.

Although C-band satellites are spaced 2 degrees apart, high-power Ku-band satellites are spaced .5 degrees apart, and transmit at 120 or more watts of power.

Because of their lower frequencies and transmitter power capabilities, C-band satellites require a large receiving dish, anywhere from 6 to 10 feet in diameter. The higher power of Ku-band satellites enables them to broadcast to a compact 18-inch-diameter dish.

The satellite system

A satellite system is made up of three basic elements:

- ☐ An uplink facility, which beams programming signals to satellites orbiting over the equator.
- ☐ A satellite that receives the signals and retransmits them back to earth.
- ☐ A receiving station on earth that includes the satellite dish.

The picture and sound information originating from a studio or broadcast facility is first sent to an uplink site, where it is processed and combined with other signals for transmission on microwave frequencies. Next, a large uplink dish concentrates these outgoing microwave signals and beams them up to a satellite located 22,247 miles above the equator. The satellite's receiving antenna captures the incoming signals and sends them to a receiver for further processing. These signals, which contain the original picture and sound information, are converted to another group of microwave frequencies, then sent to an amplifier for transmission back to earth. This whole receiver/transmitter package is called a transponder. The outgoing signals from the transponder are then reflected off a transmitting antenna, which focuses the microwaves into a beam of energy that is directed toward the earth. A satellite dish on the ground collects the microwave energy containing the original picture and sound information, and focuses that energy into a low noise block converter or LNB. The LNB amplifies and converts the microwave signals to yet another, lower group of frequencies that can be sent via conventional coaxial cable to a satellite receiver-decoder inside the user's house. The receiver tunes the individual transponders and converts the original picture and sound information into video and audio signals that can be viewed on a conventional television monitor and stereo system.

RCA's DSS system

The RCA DSS system is a direct broadcast satellite system that enables millions of viewers to receive many channels of high-quality digital video programs from anywhere in the continental United States. The complete system transports digital data, video and audio to the customer's home via high powered Ku-band satellites. The program provider sends its program material to an uplink site where the signal is digitally encoded. The uplink site compresses the video and audio, encrypts the video and formats the information into data "packets." The signal is transmitted to the DBS satellites orbiting the earth above the equator. The signal is then relayed back to earth and decoded by the customer's receiver. The receiver connects to the customer's phone line and communicates with the subscription service computer providing billing information as shown in Fig. 9-1 on the following page.

Technical overview

Let's now take a brief technical overview of the total DSS satellite system.

Uplink

The DSS system transports digital data, video, and audio to the customer's home via a high-powered Ku-band satellite. The program provider sends its program material to the uplink site, where the signal is digitally encoded. The "uplink" is the portion of the signal transmitted from the earth to the satellite. The uplink site compresses the video and audio, encrypts the video, and formats the information into data "packets" that are transmitted. The signal is transmitted to a satellite, where it is relayed back to the earth and decoded by the customer's receiver.

MPEG2 compression

The video and audio signals are transmitted as digital signals instead of conventional analog signals. The amount of data required to code all the video and audio information would require a transfer rate well

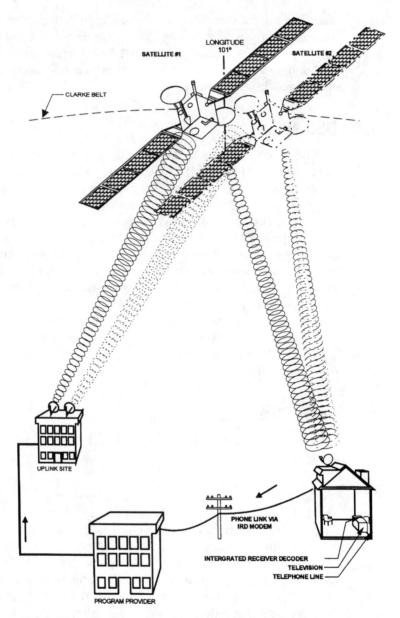

■ 9-1 *The complete Digital Satellite System.* Thomson Consumer Electronics

into the hundreds of Mbps (megabits per second). This is too large and impractical a data rate to be processed in a cost effective way with current hardware. In order to minimize the data transfer rate, the data is compressed using MPEG2 compression. MPEG (Motion Pictures Expert Group) is an organization that has developed a specification for transportation of moving images over communications data networks. Fundamentally, the system is based on the principle that images contain a lot of redundancy from one frame of video to another—the background stays the same for many frames at a time. Compression is accomplished by predicting motion that occurs from one frame of video to the next, and transmitting motion vectors and background information. By coding only the motion and background difference instead of the entire frame of video information, the effective video data rate can be reduced from hundreds of Mbps to an average of 3 to 6 Mbps. This data rate is dynamic and will change depending on the amount of motion occurring in the video.

In addition to MPEG2 video compression, MPEG2 audio compression is also used to reduce the audio rate. Audio compression is accomplished by eliminating soft sounds that are near loud sounds in the frequency domain. The compressed audio data rate can vary from 56 Kbs (Kilobits per second) on mono signals to 384 Kbps on stereo signals.

Data encryption

To prevent unauthorized signal reception, the video signal is encrypted (scrambled) at the uplink site. A secure encryption "algorithm" formula known as the Digital Encryption Standard (DES) is used to encode the video information. The keys for decoding the data are transmitted in the data packets. The customer's access card decrypts the keys, which allows the receiver to decode the data. When an access card is activated in a receiver for the first time, the serial number of the receiver is encoded on the access card. This prevents the access card from activating any other receiver except the one in which it was initially authorized. The receiver will not function with the access card removed.

Data packets

The program information is completely digital and is transmitted in data packets. This concept is identical to that of data transferred by a computer over a modem. The five types of data packets used by the system are video, audio, CA, PC-compatible serial data, and program guide packets. Video and audio packets contain the visual and audio information of the program. The CA (Conditional Access) packet contains information that is addressed to individual receivers. This includes customer E-Mail, access card activation information, and which channels the receiver is authorized to decode. PC-compatible serial data packets can obtain any form of data the program provider wants to transmit, such as stock reports or software. The Program Guide maps the channel numbers to transponders and SCIDs (more on this later). It also gives the customer TV program listing information.

A typical uplink configuration, shown in Fig. 9-2, is for one transponder. In the past, a single transponder was used for a single satellite channel. With digital signals, more than one satellite channel can be sent on the same transponder. The example shows

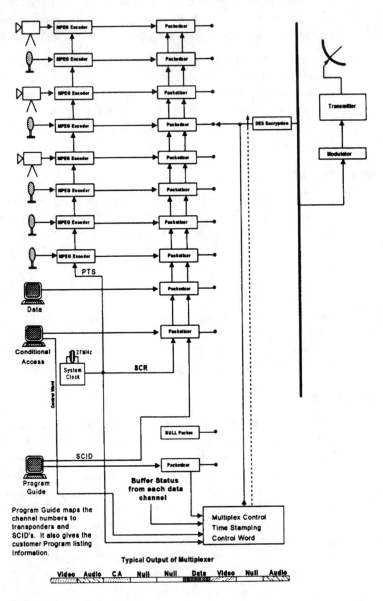

9-2 *The uplink setup.* Thomson Consumer Electronics

three video channels, five stereo audio channels (one for each video channel plus two extra for other services, such as audio in a second language), and a PC-compatible data channel. Audio and video signals from the program provider are encoded and converted to data packets. The configurations can vary depending on the type of programming. The data packets are then multiplexed into serial data and sent to the transmitter.

Each data packet is 147 bytes long. The first two bytes (a byte is made up of eight bits) of information contain the SCID and flags. The SCID (Service Channel ID) is a unique 12-bit number that uniquely identifies the packet's data channel. The flags are made up of four bits used primarily to control whether or not the packet is encrypted and which key to use. The third byte of information is made up of a four-bit packet type indicator and a four-bit continuity counter. The packet type identifies the packet as one of four data types. When combined with the SCID, the packet type determines how the packet is to be used. The continuity counter increments once for each packet type and SCID. The next 127 bytes of information consists of the "payload" data, which is the actual usable information sent from the program provider (refer to Fig. 9-3).

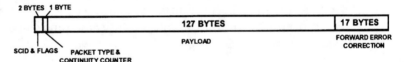

■ **9-3** *A data packet.* Thomson Consumer Electronics

Satellites

Two high-power Ku-band satellites provide the DSS signal for the receiver. The satellites are located in a geostationary orbit in the "Clarke" belt, 22,247 miles above the equator. They are positioned less than 0.5 degrees apart from each other with the center between 101 degrees west longitude. This permits a fixed antenna to be pointed at the 101-degree slot and receive signals from both satellites. The downlink frequency is in the K4 part of the Ku-band, at 12.2 to 12.7 GHz. The total transponder channel bandwidth is 24 MHz per channel, with channel spacing at 14.58 MHz. Each satellite has sixteen 120-watt transponders.

Unlike C-band satellites, which use horizontal and vertical polarization, the DSS satellites use circular polarization. The microwave energy is transmitted in a spiral-like pattern. The direction of rotation determines the type of circular polarization, as shown in

Fig. 9-4. In the DSS system, one satellite is configured for only right-hand circular polarization transponders, and the other is configured for only left-hand circular polarization transponders. This nets 32 total transponders between two satellites. These satellites have a life expectancy of 12 years (refer to Fig. 9-5).

Right Hand Circularly Polarized Wave   **Left Hand Circularly Polarized Wave**

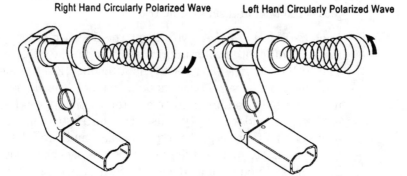

■ **9-4** *Right-hand and left-hand circular polarization.* Thomson Consumer Electronics

Although there are only 16 transponders per satellite, the channel capabilities are far greater. Using data compression and multiplexing, the two-satellites working together have the possibility of carrying over 150 conventional (non-HDTV) audio and video channels via 32 transponders.

The dish

The "dish" is an 18-inch, slightly oval-shaped Ku-band antenna. The slight oval shape is due to the 22.5 degree offset feed of the LNB (low noise block converter), as shown on the following page in Fig. 9-6. The offset feed positions the LNB out of the way so it does not block any surface area of the dish, preventing attenuation of the incoming microwave signal.

The LNB

The LNB converts the 12.2- to 12.7-GHz downlink signal from the satellites to the 950- to 1450-MHz signal required by the receiver tuner. Two types of LNBs are available: dual- and single-output. The single-output LNB has only one RF connector, while the dual output LNB has two as shown in Fig. 9-7 on the following page. The dual-output LNB can be used to feed a second receiver or some other form of distribution system. The basic package comes with the single-output LNB. The deluxe package comes with the dual-output LNB.

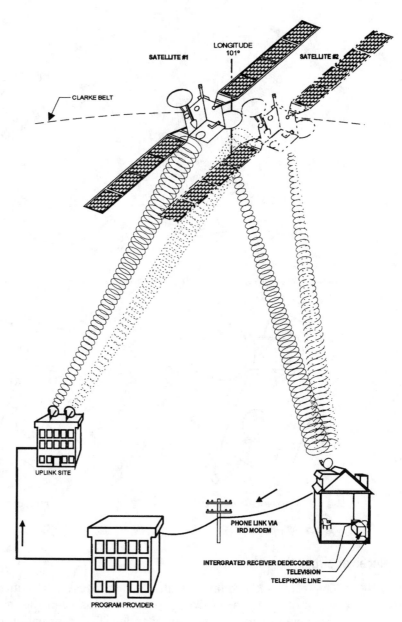

■ 9-5 *The DSS system.* Thomson Consumer Electronics

Both types of LNBs can receive both left- and right-hand polarized signals. Polarization is selected electrically via a dc voltage sent on the center conductor of the cable from the receiver. Right-hand polarization is selected with +13 volts, and left-hand polarization is selected with +17 volts.

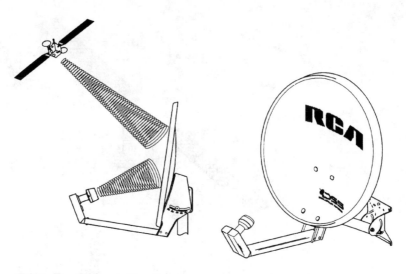

■ **9-6** *The DSS satellite dish.* Thomson Consumer Electronics

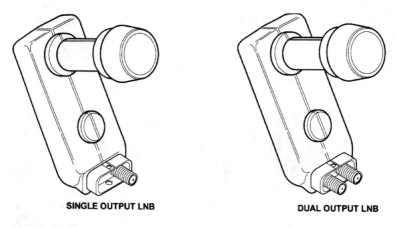

SINGLE OUTPUT LNB   **DUAL OUTPUT LNB**

■ **9-7** *Single- and dual-output LNB's.* Thomson Consumer Electronics

Receiver circuitry

The receiver is a complex digital signal processor. The amount and speed of data the receiver processes rivals even the faster personal computers on the market today. The information received from the satellite is a digital signal that is decoded and digitally processed. There are no analog signals to be found, except for those exiting the NTSC video encoder and the audio DAC (digital-to-analog converter) (refer to Fig. 9-8).

The downlink signal from the satellite is down-converted from 12.7–12.2 GHz to 950–1450 MHz by the LNB converter. The tuner

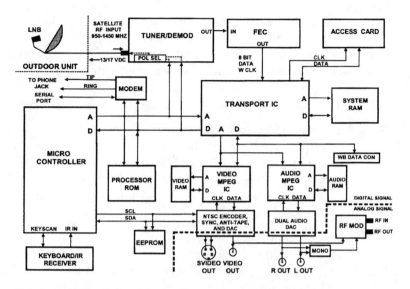

■ **9-8** *A DSS receiver block diagram.* Thomson Consumer Electronics

then isolates a single, digitally modulated 24-MHz transponder. The demodulator converts the modulated data to a digital data stream.

The data is encoded at the transmitter site by a process that enables the decoder to reassemble the data and verify and correct errors that may have occurred during transmission. This process is called Forward Error Correction (FEC). The error-corrected data is output to the transport IC via an 8-bit parallel interface.

The transport IC is the heart of the receiver data processing circuitry. Data from the FEC block is processed by the transport IC and sent to respective audio and video decoders. The microprocessor communicates with the audio and video decoders through the transport IC. The access card interface is also processed through the transport IC.

The access card receives the encrypted keys for decoding a scrambled channel from the transport IC. The access card decrypts the keys and stores them in a register in the transport IC. The transport IC uses the keys to decode the data. The access card also handles the tracking and billing of these services.

Video data is processed by the MPEG video decoder. This IC decodes the compressed video data and sends it to the NTSC encoder. The encoder converts the digital video information into NTSC analog video, which is output to the S-Video and standard composite video output jacks.

Audio data is likewise decoded by the MPEG audio decoder. The decoded 16-bit stereo audio data is sent to the dual DAC (Digital to Analog Converter), where the left and right audio channel data are separated and converted back into stereo analog audio. The audio is output to the left and right audio jacks and is also mixed together to provide a mono audio source of the RF converter.

The microprocessor receives and decodes IR remote commands and front keyboard commands. Its program software is contained in the processor ROM (read-only memory). The microprocessor controls the other digital devices of the receiver via address and data lines. It is responsible for turning on the green LED on the ON/OFF button.

The modem connects to the customer's phone line. It calls the program provider and transmits the customer's program purchases for billing purposes. The modem operates at 1,200 bps, and is controlled by the microprocessor. When the modem first attempts to dial, it sends the first number as touch-tone. If the dial tone continues after the first number, the modem switches to pulse dialing and redials the number. If the dial tone stops after the first number, the modem continues to dial the rest of the number as a touch-tone number. The modem also automatically releases the phone line if the customer picks up another phone on the same extension.

Diagnostics

The receiver contains two diagnostic test menus. The first test is a customer controlled menu that checks the signal, tuning, phone connections, and access card. The second test menu is servicing-controlled. It checks the majority of the receiver for problems.

Customer-controlled diagnostics

The customer-controlled test helps the customer during installation at any time that the receiver appears not to function properly. You have the following options in customer-controlled testing:

☐ Signal test: This test checks the value of the error bit number and the error rate to determine if the antenna connections are properly installed.

☐ Tuning test: This test checks to ensure that a transponder can be tuned. The test is considered successful and this part of the test is halted if proper tuning occurs on 1 of the 32 transponders.

204

☐ Phone test: The phone test checks for a dial tone and performs an internal loopback test.

☐ Access card test: This test sends a message to the access card and checks for a valid reply.

The response for all tests will be an "OK" display or an appropriate message informing the customer the general area of the problem.

To enter the system test feature:

1. Select "Options" from the "DSS Main Menu" as shown in Fig. 9-9.

2. Select "Setup" from the "Options" menu as shown in Fig. 9-10.

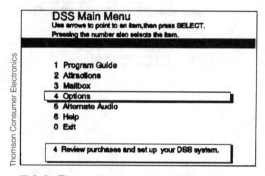

■ **9-9** *The main menu.*

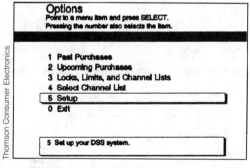

■ **9-10** *The Options menu.*

3. Select "System Test" from the "Setup" menu as illustrated in Fig. 9-11 on the following page.

4. Select "Test" from the "System Test" menu as shown in Fig. 9-12 on the following page.

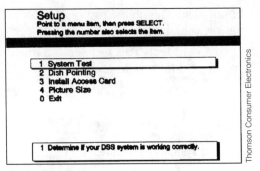

9-11 *The Setup menu.*

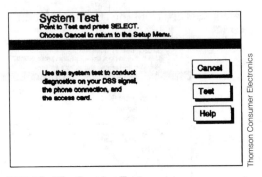

9-12 *The System Test menu.*

The system test results are displayed automatically when the test is complete. The two screens in Fig. 9-13 will show you whether the receiver passed or failed the test. If the access card passes the test, the access card ID number will be displayed in the window.

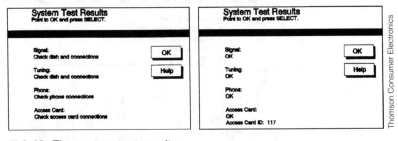

9-13 *The system test results.*

Servicer-controlled diagnostics

The servicer-controlled test provides a more in-depth analysis of the receiver for proper operation. The test pattern checks all possible connections between components as a troubleshooting aid. The following information is provided to the servicer:

1. IRD serial number

2. Demodulator vendor and version number

3. Signal strength

4. ROM checksum results

5. SRAM test results

6. V-Ram test results

7. Telco callback results

8. Verifier Version

9. Access card Test and Serial Number

10. IRD ROM version

11. EEprom test results

The response for all tests will indicate the test was successful or not successful. In addition, this menu will allow entry into the phone prefix menu so the installer can set up a one-digit phone prefix.

To enter the service test feature

Simultaneously press the front panel "TV/DSS" button and the down arrow button. A screen will appear as shown in Fig. 9-14. The test results are automatically displayed after the test is complete. The servicer is given the option to exit or run the test again.

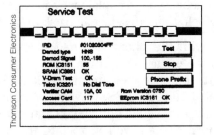
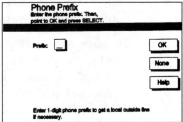

■ **9-14** *The Service Test menu.*

Also included in the Service Menu are previsions for testing the modem and setting a single-digit prefix number. During the service test, the modem will dial the phone number that appears in the boxes at the top of the test menu. The phone number can be changed by using the down arrow keys on the remote control or receiver to move the cursor past the Prefix prompt to the number boxes. Once the boxes are selected, the number can be entered or changed with the number keys on the remote or by using the up/down keys on the remote receiver. The prefix can be changed by selecting "Phone Prefix" on the display and changing the number with the number keys on the remote control or by using the arrow keys on the remote control and front panel. These front panel buttons are shown in Fig. 9-15.

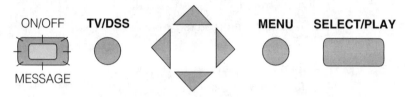

ON/OFF   TV/DSS   MENU   SELECT/PLAY

MESSAGE

■ **9-15** *The front panel button layout.* Thomson Consumer Electronics

Site survey information

The purpose of the site survey is to plan the installation of the digital satellite system. This planning includes the locations of the dish, receiver, and routing of cables. This helps to determine the tool and hardware requirements for the installation, plus it identifies any potential problems.

While performing the site survey, one of the most important things to do is work with the customer. An installation requires drilling holes and routing cables through their home. The customer may have specific locations where they want this done. Involving them in the site survey provides you an opportunity to learn what these preferences are. It also enables you to explain why suggested locations may not work and help them to select alternative locations.

The site survey starts with the phone call to the customer to arrange the date and time of the installation. This phone call is your initial contact with the customer and most likely the only contact before the installation of the DSS system. If possible, try to determine as much about the installation as you can during this phone call. Ask questions to help you anticipate the tool and equipment needs for the installation.

These questions should include:

☐ How many televisions will be connected to the system?

☐ Will an audio system be connected?

☐ Are there any preferences for dish and receiver locations?

The more of this information you can obtain before the installation appointment, the better your chances of having the right tools and mounting hardware needed to complete the installation, and the less time it will take.

Another question to ask the customer before the installation appointment is if there are any codes, covenants, regulations, and restrictions pertaining to the installation of the DSS dish. Knowing these covenants is the responsibility of the customer, but any input you can offer may reduce any problem that may arise.

The next step of the site survey occurs at the customer's home at the time of the installation. At this point, the installer and customer should work together to determine the details of the installation. These details should include the following:

☐ A location for the dish.

☐ A dish mounting system (horizontal, vertical, chimney, or pole).

☐ A route for the cables that run from the dish to the receiver.

☐ How to connect the digital satellite system to the customer's audio/video components.

☐ Evaluation of off-air antenna and/or cable solutions.

DSS dealer demonstration

The photo in Fig. 9-16 is a dealer retail demonstration set-up of the DSS system that gives consumers an opportunity to "test

■ **9-16** *A DSS demonstration unit for retail sales (USSB).*

drive" the system and experience the delights of digital programming and digital video/audio viewing.

DSS troubleshooting

To make troubleshooting easier, the DSS receiver has a diagnostic system built into it. This system makes several checks of key areas in the receiver to determine their status. The tests made are in the following areas:

☐ Signal

☐ Tuning

☐ Telephone

☐ Access card

Once the receiver completes these checks, the on-screen menus displays the results of each check. For example, if the tuner checks good, the on-screen display will indicate it is OK. If it checks bad, the receiver displays a failure message. For example, it may say to check the dish and cables. Using the diagnostics menus, it is possible to localize most problems to either the receiver or other components of the Digital Satellite System. The diagnostic menu is shown in Fig. 9-17.

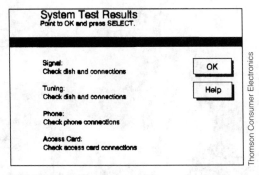

System Test Results
Point to OK and press SELECT.

Signal:
Check dish and connections

Tuning:
Check dish and connections

Phone:
Check phone connections

Access Card:
Check access card connections

OK

Help

Thomson Consumer Electronics

■ **9-17** *The Consumer Diagnostics menu.*

Troubleshooting procedures

One of the first things to do when troubleshooting a suspected digital satellite system problem is to confirm the connections between the receiver and television. When the receiver is on, it produces on-screen displays. If the television is on an operates, but will not display the on-screen menus from the DSS receiver, either the receiver is not functioning or the cable connections between the television and the receiver are bad. To determine which of the two is the problem, check out or substitute the coax cables. If this does not solve the problem, then replace the receiver.

If the on-screen menus are visible on the television, use the receiver's diagnostics menus to help determine the problem. To do this, use the front panel controls or the IR transmitter of the receiver to place it in the customer test mode. Once in the test mode, the receiver runs four tests. These tests were explained in the technical overview. If one of these tests fail, use the following procedures to determine the cause of the failure:

Signal test

If the signal test fails the display will indicate "Check Dish and Coax Connections." When this occurs, make the following checks:

Test and check procedures

1. With a voltmeter, check the dc voltage applied to the LNB. To do this, remove the coax cable from the LNB and measure between the center conductor and shield of the cable. This voltage should be between 12 and 18 volts. If correct, go onto step 2. If not suspect the LNB cable or receiver.

2. Verify that the dish is pointed at the DSS satellites. To confirm this, use the signal strength meter in the DSS alignment menu. This menu has two indicators that are important to troubleshooting. These are signal strength and lock. The signal strength meter indicates the amount of signal received by the receiver. The lock indicator indicates whether the received signal is the DSS signal. If the receiver is locked to a DSS signal but signal strength is low, chances are good that there is a dish pointing error or a defective LNB. If the receiver is receiving a strong signal but not locked, either the dish is pointing at the wrong satellite or the receiver is defective. If the menu shows both the signal strength low and the lock indicator as unlocked, a defective receiver, LNB, or an error in dish pointing could be the problem.

3. Substitute the LNB. Once the LNB is substituted, once again check the operation of the receiver. Do this with the customer diagnostics menu. If the diagnostics menu signal strength indicator checks OK, the original LNB was defective. If the signal strength indicator still does not check okay, suspect the receiver.

Step 1 of this procedure verifies that the LNB is receiving power from the receiver. If not, the next check would be at the "satellite in" jack at the rear of the receiver. To do this, unscrew the coax cable from this jack and measure the voltage between the center of the jack and the shield. If this voltage measures between 12 and 18 volts, suspect a short in the LNB cable. If the LNB voltage is missing, suspect a defective receiver.

Step 2 verifies that the dish is pointing at the satellite. To confirm this, use the signal strength meter in the receiver's alignment menu. This menu has two indicators that are important to troubleshooting. These are signal strength and lock. The signal strength meter indicates the amount of signal received by the receiver. The lock indicator indicates whether the received signal is the DSS signal. If the receiver is locked to a DSS signal but the signal strength is low, chances are good that there is a dish-pointing error or a defective LNB. If the receiver is receiving a strong signal but is not locked, either the dish is pointing at the wrong satellite or the receiver is defective. If both the signal strength is low and the receiver is unlocked, a defective receiver, LNB, or an error in dish pointing could be the problem. Refer to the consumer diagnostics menu in Fig. 9-18.

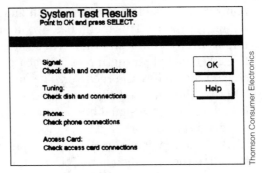

■ **9-18** *The Dealer Diagnostics menu.*

First, confirm that the dish is pointed at the DSS satellites. Find the azimuth and elevation coordinates from the dish to the satellite. Then use a compass to verify the azimuth. Verify the elevation by checking the alignment of the indicator on the LNB support arm. When checking the dish alignment, be concerned about any bent pieces on the dish, plus any shifting of the mounting foot on the mounting surface. If nothing looks out of place, use a level or plumb line to confirm that the mounting foot and mast are level. If the dish is pointed correctly, then go onto the next step.

In step 3, the LNB is substituted to determine if it is defective. Once the LNB is substituted, the operation of the receiver should be rechecked. Do this by activating the alignment screen. If the signal strength is good and locked, the original LNB was defective. If the signal still does not check out, suspect the receiver or the LNB cable.

Tuning

The tuner check indicates whether the receiver's tuner is tuning to a satellite transponder. If the signal strength is good but the tuning test fails, suspect a defective receiver or LNB.

Phone

During the phone portion of the self test, the receiver checks for a dial tone and performs an internal loopback test. If the phone test detects a problem, it will put the message "Check phone connections" on the screen. If the phone check detects no problems, the message OK will appear on the screen.

If the phone check fails, you must confirm the operation of the telephone line connected to the receiver. Do this by unplugging the receiver and connecting a known-good telephone in the receiver's place. If the telephone works, suspect a defective receiver. If the telephone does not work, suspect a defective connection or a bad cable between the junction box and the receiver's connecting point.

Access card

During this test, the receiver sends a code to the access card and looks for a reply. If the card replies correctly, the card is OK. If the card does not answer back or answers incorrectly, the card is malfunctioning.

DSS dish installation tips

A course alignment of the satellite antenna to locate the satellite can be accomplished by using the following information. The map in Fig. 9-19 on the following page shows latitude and longitude lines running through the United States. The additional curved lines on the map represent the magnetic variation of the earth's magnetic field. The chart in Fig. 9-20 on the following page shows the azimuth and elevation to be used when pointing the dish for a given location. Both the true north and magnetic north azimuths are provided.

Because the earth's geographical north pole is not the same as the magnetic north pole, magnetic north and true north readings are

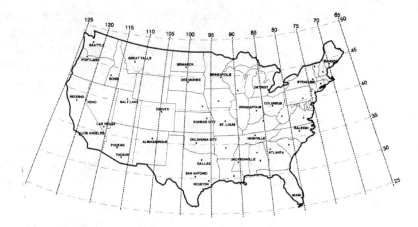

■ **9-19** *A latitude and longitude map of the United States.* Thomson Consumer Electronics

not the same. A compass, being a magnetic device, will point to magnetic north, not true north. To point the dish in the right direction, the magnetic variation must be considered when calculating the azimuth. The charts show the magnetic variation, true north azimuth and the magnetic north azimuth. The magnetic north is the reading used when finding the azimuth with a compass.

To point the dish with the map chart:

1. Find the approximate location of the installation site on the map.
2. Locate the nearest latitude (horizontal) line.
3. Locate the nearest longitude (horizontal) line.
4. Find the row in the chart with the same latitude and longitude as found in step 2 and 3.
5. Use the magnetic north azimuth and its associated elevation to point the dish in the right direction.

For example, if the dish were being installed in Houston, Texas, the latitude would be approximately 30 degrees and the longitude approximately 95 degrees west. The magnetic north azimuth would be approximately 186 degrees, with an elevation of approximately 54.4 degrees. Although these values are only approximate, they will get the dish pointed in the general direction so a fine alignment can be made using the pointing menus in the receiver.

The DSS receiver also calculates the azimuth and elevation to the DSS satellites from the installation site. It does this with two methods. One method uses the zip code of the installation site, while the other method uses the installation site's azimuth and eleva-

SATELLITE 101º WEST					
LATITUDE	LONGITUDE	*MAG. DECL.	TRUE NORTH AZ	*MAG. NORTH AZ	ELEV.
25	80	+2	222	224	52.5
25	85	0	214	214	55.7
25	90	-4	205	201	58.3
25	95	-6	194	188	60.0
25	100	-8	182	174	60.8
25	105	-9	171	163	60.4
25	110	-10	159	149	59.1
25	115	-12	149	137	56.8
30	80	+3	218	221	48.2
30	85	0	210	210	50.9
30	90	-3	201	198	53.0
30	95	-6	192	186	54.4
30	100	-8	182	174	56.2
30	105	-10	172	162	54.8
30	110	-11	162	151	53.7
30	115	-13	154	141	51.8
35	75	+7	220	227	40.9
35	80	+5	214	219	43.6
35	85	+1	207	208	45.9
35	90	+3	199	196	47.7
35	95	-6	190	184	48.9
35	100	-9	182	173	49.3
35	105	-11	173	154	49.1
35	110	-13	165	152	48.2
35	115	-14	157	143	46.7
35	120	-15	149	134	44.6
40	70	+16	223	239	33.8
40	75	+11	217	228	36.6
40	80	+7	211	218	38.9
40	85	+3	204	207	40.9
40	90	-2	197	195	42.4
40	95	-6	189	183	43.3
40	100	-9	182	172	43.7
40	105	-12	174	162	43.6
40	110	-14	166	152	42.8
40	115	-16	159	143	41.5
40	120	-17	152	135	39.8

■ **9-20** *DSS azimuth and elevation information.*

tion. Before you can use either of these methods, you must first connect your satellite receiver to a TV set. Once the TV is connected to the receiver, you can bring up the dish-pointing information on the menus.

DSS installation certification test

Select the best answer to the following questions and fill in the corresponding circle on the answer sheet shown in Fig. 9-21.

DIGITAL SATELLITE SYSTEM

Answer Sheet

Installer Technician Name: _____
(Please Print)

Installer Technician Home Address: _____
(Street)   (City)   (State)   (Zip)

Installer Company Name: _____
(Please Print)

Installer Company Address: _____
(Street)   (City)   (State)   (Zip)

Installer Company Phone Number: (   )_____FAX ( )_____

6-digit Installer Authorization Number: ☐☐☐☐☐☐
(This is the Installer Authorization number assigned to your company by Thomson.)

Installer Technician signature:_____ Test Date: _____

	A B C D		A B C D
1.	O O O O	14.	O O O O
2.	O O O O	15.	O O O O
3.	O O O O	16.	O O O O
4.	O O O O	17.	O O O O
5.	O O O O	18.	O O O O
6.	O O O O	19.	O O O O
7.	O O O O	20.	O O O O
8.	O O O O	21.	O O O O
9.	O O O O	22.	O O O O
10.	O O O O	23.	O O O O
11.	O O O O	24.	O O O O
12.	O O O O	25.	O O O O
13.	O O O O	26.	O O O O

■ **9-21** *The DSS answer sheet.* Thomson Consumer Electronics

1. What is the typical wattage of a DSS satellite transponder?
   A. 10
   B. 75
   C. 120
   D. 150

216

2. What are the three basic elements of a satellite system?
   A. Uplink, downlink, and LNB.
   B. Uplink, satellite, and receiver.
   C. Modulator, demodulator, and encoder.
   D. Dish, LNB, and coaxial cable.

3. What is an LNB?
   A. A low-noise blower used to cool the receiver.
   B. Latitude navigator block.
   C. Low noise block converter.
   D. Line noise blocker.

4. Where are the DBS satellites approximately located?
   A. 101 degrees east.
   B. 105 degrees west.
   C. 102 degrees north.
   D. 102 degrees west.

5. What is the difference between the basic package LNB and the deluxe package LNB?
   A. The basic package LNB is lighter in color than the deluxe package LNB.
   B. They are identical.
   C. The deluxe package LNB has 2 outputs and the basic package LNB has only 1.
   D. They are the same except for the mounting bracket.

6. How does the program provider know what programs the customer has watched?
   A. The receiver periodically calls the program provider computer via telephone.
   B. The dish transmits data backwards on the microwave carrier to the program provider.
   C. The customer calls a 1-800 number monthly and tells the operator what they have watched.
   D. The program provider does not know what the customer has been watching.

7. The DSS signal is what type of signal?
   A. Short-wave analog.
   B. Digital.
   C. Microwave analog.
   D. Pulse-width-modulated.

8. How many channels are broadcast per transponder?
   A. One channel per transponder.
   B. More than one channel per transponder using data compression and multiplexing.

217

C. Two channels only—1 main and 1 backup.

D. More than 150 channels on one transponder.

9. What will the receiver do if the access card is removed?

A. Work normally.

B. Cease to decode audio and video.

C. Have no second-channel sound.

D. The remote control will not function.

10. Once activated, why will the access card work in only one receiver?

A. Once the card is inserted, it cannot be removed without first erasing its contents.

B. Each card will only fit the receiver in which it was originally inserted.

C. The access card will work in any receiver in which it is inserted.

D. The card is coded to only work in the receiver in which it was activated.

11. How are right-hand and left-hand circular polarization selected on the LNB?

A. By a motor rotating the LNB clockwise and counter-clockwise.

B. By a dc voltage sent on the center conductor of the LNB cable.

C. By an RF remote control signal sent to the LNB receiver.

D. By a mechanical switch on the outside of the LNB.

12. What are the correct voltages for right-hand and left-hand polarization?

A. 5 volts for right-hand and 0 volts for left.

B. 10 volts for left and 0 volts for right.

C. There is no voltage, because the polarization is selected by a mechanical switch on the LNB.

D. +13 volts for right hand and +17 volts for left.

13. What will cause the modem to automatically hang up?

A. When it dials the wrong number.

B. If it detects an off-hook condition of another phone on the same line.

C. When the customer presses the "Hang-up" button on the remote unit.

D. When it detects an incoming call.

14. What do the customer-controlled diagnostics tell the installer?

A. It acknowledges whether a signal can be received and tuned, if the modem works and if the access card is OK.

B. It tells if the transport IC is functioning correctly.

C. It checks to make sure the TV is connected properly to the receiver.

D. It tells whether or not there has been a power failure.

15. What needs to be grounded on the dish?
    A. The dish and LNB cable.
    B. The wall and RF line out of the receiver.
    C. The telephone wall plate.
    D. Only the LNB.

16. What would the picture look like if the dish were not pointed at the DSS satellite during initial alignment?
    A. Weak picture with sparkles/snow.
    B. No picture at all—blank screen.
    C. All snow, no picture at all.
    D. Green picture.

17. When using the audio tone to align the dish, at what point does the tone switch from short bursts to a continuous tone?
    A. When the receiver has finished warming up.
    B. When you point the dish at the sun.
    C. When the DSS signal is acquired.
    D. When time has run out on the subscription.

18. What is the purpose of the coaxial cable ground block?
    A. It provides an easy place to splice the cable.
    B. It is a safety measure to help prevent damage and/or electrical shock from lightning or other similar electrical hazards.
    C. Required by most home-owner associations to make the installation look like cable TV.
    D. Both A and C are correct.

19. What is the purpose of the site survey?
    A. To get to know the customer.
    B. To plan the details of the DSS installation with the customer.
    C. To see what color the house is so the dish can be painted to match.
    D. To check and see if the neighbor has any objections to the installation.

20. How is the system activated?
    A. An installer will call the program provider(s) and give them the necessary information to activate the system.
    B. The installer will input a secret code with the remote control to activate the system.

219

C. The receiver comes already set up and ready to go.

D. The customer mails in an order form and within six to eight weeks the program provider will activate the box via satellite.

21. What is the last step in the installation process?

A. Routing the cable from the LNB to the ground block.

B. Pointing the dish.

C. Spending at least 20 minutes educating the customer about system operation.

D. Connecting the phone cable.

22. What type of cable must be used for the LNB?

A. RG-59

B. RG-58

C. RG-6

D. RG-3

23. What is the purpose of the drip loop in the cable?

A. It prevents moisture from entering the house and connections.

B. It makes the installation look professional.

C. It prevents the RF signal from being too strong going into the house.

D. It takes up the extra slack in the cable.

24. What is the most important aspect to remember when installing a phone cable?

A. Make sure your hands are dry.

B. Follow the color code.

C. Use the correct electrical tape.

D. Do not nick the wire when stripping it.

25. When doing a line-of-sight survey, what factors would not disqualify a sight?

A. Trees.

B. Cloud cover.

C. Buildings.

D. Mountains.

26. What must the mounting foot be secured to when mounting to a wood surface or roof?

A. A wall stud or rafter.

B. Stucco.

C. Aluminum siding.

D. Rain gutter.

Detailed DSS receiver, coax cable, and telephone installation information

10

THIS CHAPTER BEGINS WITH DETAILED INSTRUCTIONS FOR installing the LNB coax cable between the DSS dish and the satellite receiver. It will tell you how to make the proper connections on back of the satellite receiver and how to accomplish proper grounding of the DSS dish.

You will find detailed information for installing the telephone cable and modular cable jack, as well as wireless phone adaptor information. Various inputs and outputs from the DSS receiver are then explained. The standard and deluxe satellite connections are also illustrated and explained.

The chapter concludes with more dish alignment and review notes.

Installing the LNB cable

Depending on the installation site and the type of system installed, there may be up to three cables run into the home during installation. One cable is from the dish, and carries the LNB signal to the receiver. The second cable is a telephone cable. If you are installing the deluxe system, two cables carry LNB signals to two different receivers, while the third cable is a telephone cable. The following explanation describes the installation of a basic system. The installation of an advanced system is the same, except for the addition of one LNB cable. A description of the telephone cable installation appears later in this chapter.

The LNB cable carries signals from the LNB to the receiver. These signals are in the frequency range of 950 to 1450 MHz. It requires a cable with low signal loss to carry signals in this frequency range.

A poor quality cable may allow noise to enter the system, reducing its performance. Minimum specifications for the LNB cable are given in Fig. 10-1.

Specification	Rating	General comments
Cable type	RG6	
Impedance	75 ohms	
Shielding	Minimum double shield	Requires a minimum 100% foil shield covered with a 40% woven braid.
Outer cover	PVC	Must be suitable for both indoor and outdoor use.

■ **10-1** *LNB cable specifications.* Thomson Consumer Electronics

Due to the frequency range of the signals carried by the LNB cable, it must be RG6 coax cable. RG6 cable has the correct impedance (75 ohms) and acceptable signal losses at 950 to 1450 MHz. When selecting an RG6 coax cable for the LNB signal, select a type that is double-shielded with a 100% braid foil shield, covered with at least 40% woven braid. If you are in an area that has a lot of RF noise, a woven braid shield of higher than 40% may be required. If you are unsure of your cable specifications, ask the cable supplier for the specifications of the cable you are using.

Depending on the routing of the LNB cable, you may want to use a type that can be buried. When normal coax cable is buried, soil decays its outer cover and shortens its life. Cables that are suitable to be buried use a special outer cover that resists breakdown. Some of these cables also have a special coating on the ground shield. This coating resists corrosion caused by water, which could get in the cable. Anytime the LNB cable is buried, use the proper grade of cable. It will save you problems in the future.

The goal of any cable installations should be to protect the cable from physical damage and moisture penetration. To protect the cable from physical damage, secure it to walls or any stable surface with cable clamps. This prevents the cable from sagging and being damaged by people stepping on it or running over it with yard equipment. You can prevent moisture penetration by using weatherproof connectors or sealing any connection that is exposed to the elements. Drip loops add additional protection in the area of a connection. These loops stop moisture from traveling down the cable and entering the connection.

The LNB cable supplies the signal from the LNB to the receiver. This cable also carries the dc power to the LNB from the receiver.

222

For the LNB to receive power and operate, it is important that any splitters or amplifiers inserted in the LNB cable must pass dc.

Another important aspect of the LNB cable installation is grounding. The National Electrical Codes (Article 820-33) requires that any coax cable exposed to possible contact with lightning or power conductors must have the outer shield grounded. There also may be other rules regarding grounding in your area. It is important to consult and follow all codes and regulations in your area during installation.

The best method of grounding the outer shield of a coax cable is with a ground block. This block is a barrel connector with a means to connect a ground conductor (like a terminal or set screw). The terminal is where the outer conductor of the coax cable is connected to the grounding electrode. When installed, the ground block passes the LNB signal and dc voltages straight through it while grounding the outer conductor of the coax. Note drawing shown in Fig. 10-2.

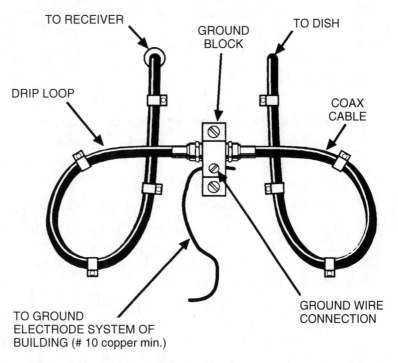

■ 10-2 *A coax cable ground block and drip loops.* Thomson Consumer Electronics

Where should the ground block be placed in the LNB cable? One factor determining the ground block location is the location of an acceptable ground electrode. The National Electrical Code specifies that the grounding conductor (the wire that runs between the ground block and ground electrode) must be in a straight line, if possible, from the ground block to the grounding electrode. Also, splices in the conductor between the electrode and grounding block are prohibited. Article 820-33 of the National Electrical Codes also states that "Where coaxial cable is exposed to lightning . . . the cable ground shall be connected to the grounding system of the building, as close to the point of cable entry as practical." This can be either inside or outside the home. Therefore, when selecting a location for the grounding block, try to make the ground conductor as short and straight as possible, and close to the cable entrance point into the home.

The National Electrical Code also has specifications for the size of the grounding conductor that connects to the ground electrode. This conductor must be at least a No. 10 copper wire or No. 8 aluminum wire. Insulation over the conductor wire is not required. As with any cable, the grounding conductor must be securely fastened to a surface to protect it from physical damage. If the conductor cannot be protected from damage, the size of the wire should be increased to withstand any physical strain placed on it.

The National Electrical Code is also specific on acceptable ground electrodes. Some possible electrodes are as follows:

☐ Grounded interior metal water piping (cold water).

☐ Ground rod (must be driven at least eight feet into the ground).

☐ A grounded metallic service raceway.

☐ A grounded electrical service equipment enclosure.

There are also specific instructions for attaching the grounding conductor to grounding electrodes. For more details on this, consult the *National Electrical Code Handbook* and any active codes that pertains to your area.

Use the following steps to install the LNB cable:

1. Determine the location of the ground block. Remember to position this block to allow a short and straight route for the ground conductor to the grounding electrode.

2. With two screws, secure the grounding block to a stable mounting surface.

3. Connect one end of the LNB cable to the grounding block.

224

4. Route the cable from the ground block to the dish mounting foot. Leave enough slack in the cable to form a 3" to 5" drip loop. Also, leave enough cable to reach the LNB from the mounting foot (about three feet).

5. Secure the cable with cable clips. Do not forget to form the drip loop and secure it with clips.

6. Install the ground conductor on the ground terminal of the coax ground block.

7. Route the ground conductor to the grounding electrode. Remember to secure the cable to a wall or some surface to protect it.

8. Connect the ground conductor to the grounding electrode (note drawing in Fig. 10-3).

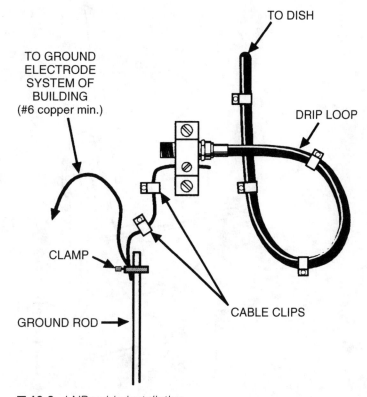

■ **10-3** *LNB cable installation.* Thomson Consumer Electronics

Attaching the LNB

The LNB attaches to the LNB support arm. To attach it, you will need the LNB mounting bolt, washer, and nut from the hardware packet. You will also need to remove the LNB from the packaging. The tools you will need are a Phillips head screwdriver and the tools required to install an "F"-type connector (you will only need this if the cable is not preterminated with an F-fitting). If the F-type connector you are using is not weatherproof, you will also need some type of coax cable sealant. This sealant prevents moisture from seeping into the LNB through the coaxial connector. Most LNB failures are due to moisture penetration. Always use some type of sealant or a weatherproof connector to protect this connection.

Use the following steps to install the LNB onto the LNB support arm:

1. Route the LNB cable through the foot, mast, and LNB support arm (refer to the drawing in Fig. 10-4).

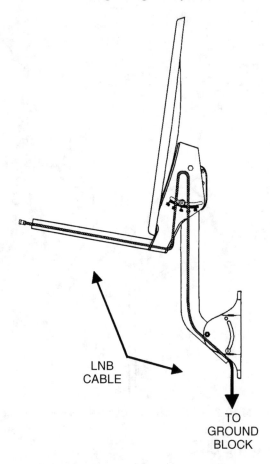

LNB
CABLE

TO
GROUND
BLOCK

■ **10-4** *Cable routing through the dish.*
Thomson Consumer Electronics

Detailed DSS receiver, coax cable, and telephone installation information

2. Install an F-type connector on the end of the cable. If the cable already has a connector on it, disregard this step.

3. Thread the coax cable onto the LNB connector (see drawing in Fig. 10-5).

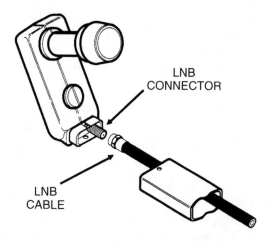

LNB
CONNECTOR

LNB
CABLE

■ **10-5** *Connecting the LNB cable to the LNB.* Thomson Consumer Electronics

4. If the connector is not weatherproof, seal it.

5. Slide the end of the LNB into the rectangular opening of the LNB support arm. Position the LNB to align it with the hole in the support arm.

6. Insert the LNB bolt into the hole on the bottom side of the LNB support arm.

7. Place the washer on the LNB nut and insert the nut into the hole on the top side of the LNB support arm.

8. Tighten the bolt with a Phillips screwdriver (note Fig. 10-6).

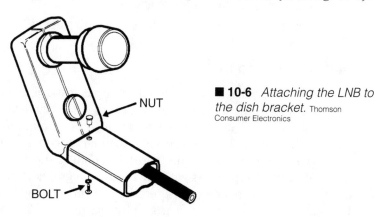

NUT

■ **10-6** *Attaching the LNB to the dish bracket.* Thomson Consumer Electronics

BOLT →

Attaching the LNB cable to the receiver

In the next step of the installation, route the LNB cable from the ground block to the receiver. In most installations, there is more than one way to get the LNB cable to the receiver from the grounding block. If the receiver is located on an outside wall, go through the wall. If the receiver is on an inside wall, use the crawl space, basement, or attic to route the cable. Whenever routing the LNB cable to the receiver, take the shortest possible path and always protect it from physical damage.

Use the following steps to install the cable from the ground block to receiver:

1. Drill a ½" hole at the access point where the cable enters the home. Before drilling, ensure that there are no wires or pipes in the area of the hole.
2. Connect one end of the cable to the ground block.
3. Form a 3" to 5" drip loop in the cable before inserting it in the access hole. Note the cable configuration shown in Fig. 10-7.
4. With cable clips, secure the drip loop and cable to the wall.
5. In the home, route the cable to the rear of the receiver. Depending on the installation site, this could be through a floor or directly to the rear of the receiver. If the cable goes straight through a wall, a wall plate may be used to dress up the access point. Remember, every attempt should be made to hide and protect the cable.
6. Connect the cable to the "SATELLITE IN" jack of the receiver (see Fig. 10-8).
7. If the connectors on the coax ground block are not weatherproof, seal them. This can be done with tape or other types of coax connector sealant.
8. Seal the access point into the home with silicone sealant.

Grounding the DSS dish

The National Electrical Code requires that masts and metal structures supporting antennas be grounded. These guidelines include the DSS dish. This grounding is similar to the ground required for the LNB cable. Basically, use a No. 10 or larger copper wire (insulation is not required on the wire) to connect the DSS foot to a grounding electrode. As with the LNB cable ground conductor, the dish ground conductor must also be securely fastened to a wall or other surface to protect it from physical damage. This conductor

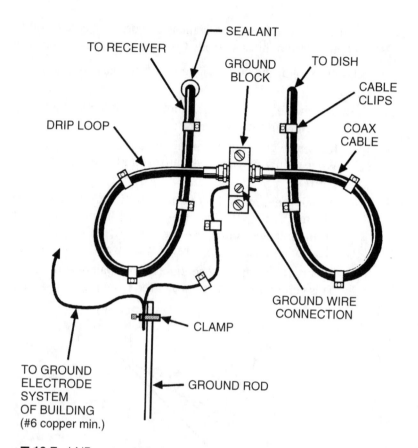

SEALANT

TO RECEIVER

GROUND
BLOCK

TO DISH

CABLE
CLIPS

COAX
CABLE

DRIP LOOP

GROUND WIRE
CONNECTION

CLAMP

TO GROUND
ELECTRODE
SYSTEM
OF BUILDING
(#6 copper min.)

GROUND ROD

■ **10-7** *LNB coax cable installation.* Thomson Consumer Electronics

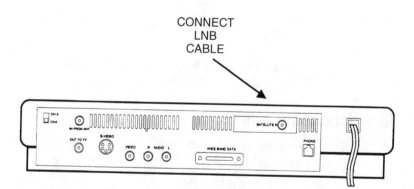

CONNECT
LNB
CABLE

■ **10-8** *Connect the LNB cable to the receiver's "Satellite In" jack.*
Thomson Consumer Electronics

must not be spliced, and it must be correctly connected to an acceptable grounding electrode (as mentioned earlier in this chapter).

Use the following steps to connect the mast to a grounding electrode:

1. Attach the grounding conductor to the mounting foot with a bolt, nut, and star washer.

2. Route the grounding conductor to the grounding electrode. While doing this, use the shortest and straightest possible path.

3. Attach the grounding conductor to the grounding electrode. The method of attaching the conductor will vary with different grounding electrodes; use the correct method for the electrode you are using (refer to drawing in Fig. 10-9).

4. Secure the grounding conductor to a wall or other surface. This conductor must be protected from physical damage.

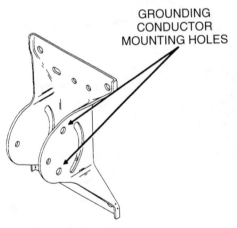

GROUNDING
CONDUCTOR
MOUNTING HOLES

■ **10-9** *Attaching the grounding conductor to the dish mounting foot.* Thomson Consumer Electronics

Telephone cable comments

There are many options available to connect the phone line to the DSS receiver. Three of the most common are:

☐ Use an existing phone jack.

☐ Install a phone jack.

☐ Use a wireless phone jack system.

To connect the receiver to an existing telephone jack, install a modular line cord between the receiver and the phone jack. The line cord has modular plugs on each end that insert into the receiver and telephone jack.

Most new homes have a telephone jack in each room. These jacks often have a single output for one telephone. Often, there is a telephone already plugged into this jack, leaving no place to connect the cable for the receiver. Should you come across this, there are several ways to increase the number of outputs from a modular phone jack.

The first of these is to use a modular duplex adapter. Another option is to replace the single wall plate with a dual type. Finally, you can use another phone jack that is located nearby. If you elect to do this, make sure to conceal the cord.

If a phone jack is not available or convenient, you may need to install a cable from a junction box or an existing phone jack. This cable can be run through a crawl space, attic, or garage to the receiver. Once the cable is at the receiver, a modular connector or a wall plate may be installed. If a modular connector is used, plug it directly into the receiver. If you install a wall plate instead of a modular plug, use a modular extension cable between the wall plate and receiver.

To install a telephone jack, it helps to understand some basics of the telephone system. The telephone company supplies a phone cable to each home. The cable connects directly to a junction block, or box, called a point of demarcation. This point divides the wiring responsibility between the customer and the telephone company. The wiring from the telephone company to one side of this junction box is the telephone company's responsibility, while the wiring connected to the other side of the box is the customer's responsibility. The customer can add additional wiring and telephones to their homes as long as they follow local and FCC guidelines and do not connect anything directly to the telephone company's side of the point of demarcation. Usually the telephone company marks or seals their side of the demarcation point to prevent improper connections.

The cable supplied to the home from the phone company normally consists of four wires. The color of these wires are red, green, yellow, and black. Homes that have only one telephone line in service use the red and green wires to carry signals. These two wires are called *ring* (red wire) and *tip* (green wire). The yellow and black wires can serve several purposes. Often they are the ring and tip wires to a second line. They may also provide power to telephone accessories. Every junction box has a terminal for each of these wires. These terminals are usually color-coded. When installing telephone cables, it is important to follow the color code.

There are cables that use a slight variation of the color code. The list in Fig. 10-10 gives you some of the various color codes. When connecting to the phone system with different cables, every attempt must be made to follow this color code.

Colors	Variation 1	Variation 2	Variation 3
Red	Blue with a white stripe	Blue with a white stripe	Blue
Green	White with a blue stripe	White with a blue stripe	White
Yellow	Orange with a white stripe	Orange with a white stripe	Orange
Black	White with an orange stripe	White with an orange stripe	White with an orange stripe
N/A	Green with a white stripe		
N/A	White with a green stripe		

■ **10-10** *Telephone wire color codes.* Thomson Consumer Electronics

The type of cable used in the phone jack installation can be critical to the performance of the system. If the cable run is short and not exposed to the weather, a typical four-conductor cable that is used for most telephone extension cables will work. If the installation requires a long cable run, select a cable that uses twisted pairs of wire. The twisting of the wires reduces the amount of noise picked up by the system. If the cable is outside and exposed to the weather, use a weatherproof cable.

NOTE: Never work with active telephone lines. When the telephone is ringing, the current carried on these lines may shock you. While working on the telephone line, disconnect the wiring at the point of demarcation. If you are unable to do this, take all the telephone handsets off the hook.

Use the following procedure to install a telephone jack:

1. Find an accessible telephone junction box (this could even be the point of demarcation). Plan the installation of the cable from that point to the receiver.

2. Remove the customer access cover on the junction box and connect the wires of the telephone cable to the terminals. Remember to follow the color code: Red wires connect to red wires, and green wires to the green ones.

3. Route the cable to the receiver. Use a crawl space, attic, or garage wherever possible. As you route this cable, secure it to walls and other supporting structures to prevent physical damage.

4. Install the modular plug or RJ11/14 type wall plate (refer to the junction box drawing in Fig. 10-11).

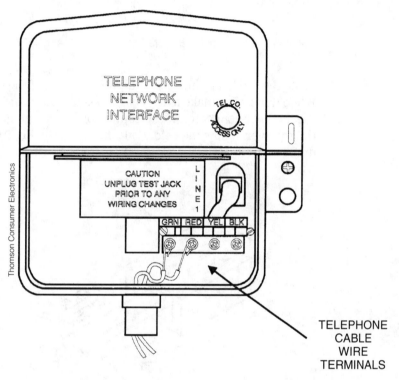

TELEPHONE
NETWORK
INTERFACE

Thomson Consumer Electronics

CAUTION
UNPLUG TEST JACK
PRIOR TO ANY
WIRING CHANGES

LINE 1

GRN RED YEL BLK

TELEPHONE
CABLE
WIRE
TERMINALS

■ **10-11**   *Telephone junction box terminals.*

Modular plug installation

Different modular crimping tools have slightly different procedures. Follow the procedure that accompanies the crimping tool you are using.

1. With the cable cut to length, trim the ends so they are straight and not diagonal.

2. Strip approximately ⅜" from the outer conductor of the cable (this is for flat, four-conductor telephone line, and may vary slightly for twist and outdoor cable). Do not strip the insulation from the four inner wires of this cable (refer to Fig. 10-12 on the following page).

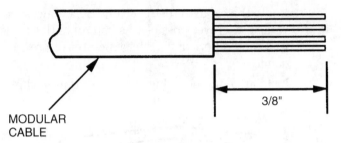

MODULAR
CABLE

3/8"

■ **10-12** *Modular plug installation steps 1 and 2.* Thomson Consumer Electronics

3. Place the modular plug onto the prepared end of the cable. While doing this, make sure the cable is oriented correctly in the modular connector. If the connector is to be plugged into the receiver, follow the color code shown in the Fig. 10-13 drawing.

234

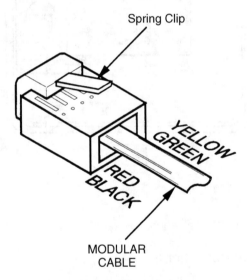

Spring Clip

YELLOW
GREEN
RED
BLACK

MODULAR
CABLE

■ **10-13** *Modular plug installation step 3.*
Thomson Consumer Electronics

4. With the wire positioned correctly in the modular connector, use a modular connector crimping tool to seat the connector on the wire as shown in Fig. 10-14.

NOTE: In most installations, the other end of the telephone cable will be handwired to a junction box. If you must install a modular

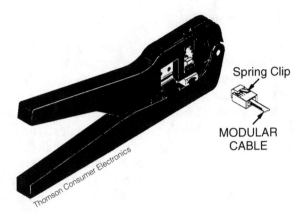

■ 10-14 *Crimping the modular plug with the crimping tool.*

connector on the other end of the cable, the orientation of the wire in the connector is different. Orientate this cable as shown in Fig. 10-15.

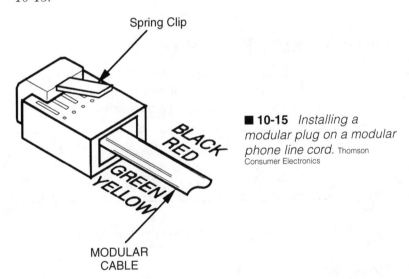

■ 10-15 *Installing a modular plug on a modular phone line cord.* Thomson Consumer Electronics

Wall plate installation

To install the wall plate, follow the instructions that accompanies the plate you purchased. Remember to follow the color code included in the wall plate instructions. Seal all entry points of the cable into the home and check to be sure it is secured and hidden from view where possible.

Wireless phone adaptor

You may need to install a wireless phone adaptor unit that uses the ac wiring within the home to connect a remote telephone (or the DSS receiver, in this case) to the telephone lines. Wireless phone adaptors usually include two pieces. One piece, called the base unit, plugs into an ac outlet near an existing phone jack. A modular telephone cable then connects between the existing phone jack and the base unit. The second piece, called the extension jack, plugs into an ac outlet near a remote phone or the DSS receiver. The telephone or DSS receiver then plugs into the extension jack. This system allows new phone jacks to be installed quickly and easily.

Installing the DSS receiver

All connections to the receiver are made at the back panel of the receiver. These connections are slightly different for the two models of receiver's.

The model DRD102RW receiver

The rear panel of the model DRD102RW receiver is shown in Fig. 10-16. Its connections are both inputs and outputs. The following is a functional list of the jack panel connections:

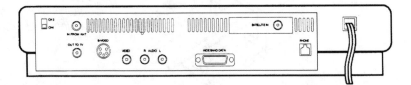

■ **10-16** *The rear panel of the DRD102RW receiver.* Thomson Consumer Electronics

Receiver inputs

☐ Phone: This jack connects via the telephone cable to the outside world. The DSS receiver uses a toll-free number to update the access card. This update usually lasts a few seconds and ensures continuous program service.

☐ Satellite in: This is the 950- to 1450-MHz input from the LNB.

☐ In from antenna: Connect either a cable TV or an outside TV antenna to this connector.

Receiver outputs

☐ Out to TV: This connector delivers the RF out (either channel 3 or 4) of the receiver. This RF can be one of two signals: One signal is RF signal applied to the connector marked "IN FROM ANT." The other signal is the output of the receiver. The TV/DSS button on the front of the receiver determines which signal is present on this jack. The RF channel output on this jack is controlled by the RF switch, and will be on either channel 3 or 4.

☐ S-video: This connector outputs an S-video signal to a compatible television or VCR. Only an S-connector can be inserted into this jack.

☐ Video: This connector outputs the video signal from the receiver, and allows you to connect the receiver to a monitor type of television or a VCR.

☐ Audio R (right) and Audio L (left): These are left and right channel audio signals from the receiver. If you are connecting the receiver to a monitor type of television, connect these to the audio inputs. You can also connect these connectors to the audio input of a stereo amplifier.

☐ Wideband data: This 15-pin connector is designed to be used in conjunction with future technology, such as high-definition TV.

237

The model DRD203RW receiver

The rear panel of the model DRD203RW receiver is shown Fig. 10-17. Its connection can be divided into inputs and outputs. The following is a functional list of the jack panel connections:

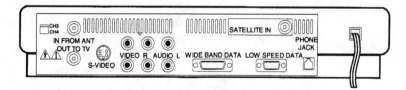

■ **10-17** *The rear panel of the DRD203RW receiver.* Thomson Consumer Electronics

Receiver inputs

☐ Phone: This jack is used for connection to the telephone cable. The DSS receiver uses a toll-free number to update the access card. This update usually lasts just a few seconds and ensures continuous program service.

☐ Satellite in: This is the 950- to 1450-MHz input from the LNB. It also carries the dc voltages from the receiver to power the LNB.

☐ In from antenna: Connect either a cable signal or an antenna to this connector.

Receiver outputs

☐ Out to TV: This is the RF out (either channel 3 or 4) of the receiver. This RF can be one of two signals: The RF signal applied to the "In from antenna" connector, or the output of the receiver. The TV/DSS button on the front of the receiver determines which signal is present on the jack. The RF channel output on this jack is controlled by the RF switch.

☐ S-video: This connector outputs an S-video signal to a compatible television or VCR. Only an S-video connector can be inserted into this jack.

☐ Video: These two connectors are bridged in the receiver. Because they are bridged, both have the same signal on them. These connectors are the video signal outputs from the receiver. One of the connectors can be connected to a monitor type of television, while the other could output video to a VCR.

☐ Audio R (right) and Audio L (left): There are two connectors for the right channel and two for the left channels. The two connectors for the right channel are bridged together and the two for the left channel are bridged together. The signals on these connectors is the right and left channel audio signals from the receiver. If you are connecting the receiver to a monitor type of television, connect one set (a set is the right and left channel) to the audio inputs. The other set can go to a VCR or stereo amplifier. If the second video output is connected to a VCR, the second set of audio cables must also be connected to the VCR.

☐ Wideband data: This 15-pin connector is designed to be used in conjunction with future technology, such as high-definition TV.

☐ Low-speed data: This 9-pin serial interface is designed to be used in conjunction with data services such as the Internet.

System connections

Figure 10-18 shows a basic connection system that will work with both the DRD102RW and DRD203RW receivers. The connection system works with a standard television set using RF cables. The TV/DSS button on the front for the receiver switches the RF signal applied to the television between the antenna (or cable) signal and the signal from the receiver.

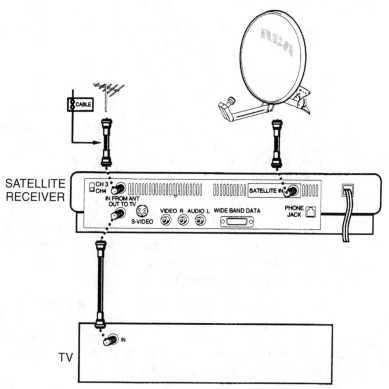

■**10-18** *Basic connects for either the DRD102RW or DRD203RW receivers.* Thomson Consumer Electronics

If the television is capable of receiving video and audio signals from the receiver, add the video and audio cables. To watch the satellite signal select the monitor inputs to the television.

To connect a stereo amplifier to this system, there are several options. If you are using a standard television, connect a stereo audio cable to the Audio R and Audio L outputs of the receiver and to the inputs of the amplifier. If your television set is a monitor and equipped with audio output jacks, then use a stereo audio cable to connect the television's audio output to the amplifier.

Figure 10-19 illustrates how a VCR is incorporated into the connection system. This system is shown using the DRD102RW receiver. This connection system applies the RF signal from an antenna or cable to the RF input of the receiver. The RF output of the receiver is sent to the VCR antenna in connector. The Out to TV connector of the VCR is connected to the antenna in connector of the television receiver. These RF connections enable the VCR to record either the antenna's or the satellite's signal. This is controlled with the TV/DSS button on the front of the receiver.

Better picture quality is achieved by using the audio video connections. To do this, connect a video cable from the video jack of the receiver to the Video in jack of the VCR. Also connect a stereo audio cable to the Audio L and Audio R jacks of the receiver. Connect the other end of the audio cable to the audio in connector of the VCR. To record the satellite signal, switch the VCR to a line input. This system also enables a satellite signal to be recorded while watching a program from the antenna or cable. To do this, place the receiver in the TV mode with the TV/DSS button. With the receiver and television's TV modes, the signal on the antenna cable will be routed to the television for viewing. At the same time, the receiver signal is applied to the audio and video input of the VCR, which can be recording it. If the connection system shown in Fig. 10-19 is followed, recording one program while watching another can be done with switches on the receiver and VCR without changing how the system is connected.

The set-up in Fig. 10-20 on the following page is a connection system that can be used with the DRD203RW. Because there are two audio video jacks on the receiver, a video and audio signal can be supplied to a second VCR or applied directly to the second input of the monitor or a stereo amplifier.

Pointing the dish

This step of the installation accurately points the dish at the satellites. This alignment is critical to the performance of the system. When the dish is pointed directly at the satellites, the receiver receives a strong signal. Any dropouts in this signal would be small, and could be corrected in the receiver. If the dish is not positioned properly, the signal will be weak and the number of dropouts then increases. Depending on the size and number of these dropouts, the receiver may not be able to correct for them and the picture will be blank or freeze frame until a good signal is once again received. This would become worse on cloudy or windy days.

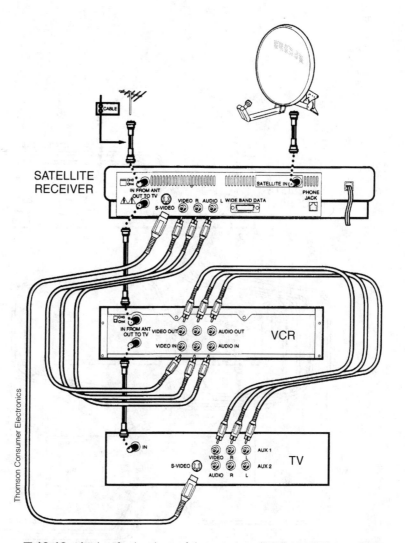

■ **10-19** *A plan for hookup of the receiver (DRD102RW) to a VCR and television receiver.*

A significant difference between an analog satellite system and the Digital Satellite System is the dish alignment process. When aligning an analog satellite system, the dish could be adjusted while looking at the television picture. When you were close to a satellite, the picture would appear with sparkles. The closer you would come, the clearer the picture became. With the Digital Satellite System, this does not happen. Instead, as you move the dish towards the satellites, the picture remains blank until a signal is received. Then the picture suddenly appears. This picture will not have black or white sparkles in it, even though the dish may not be

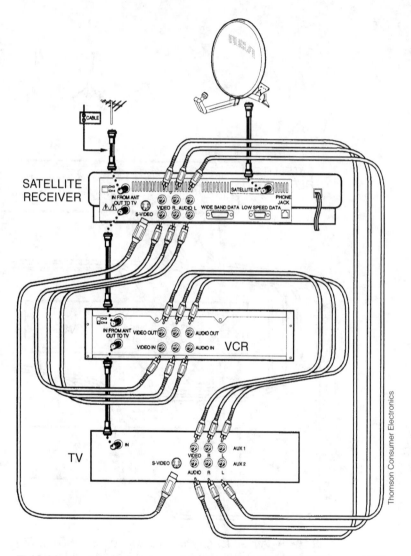

SATELLITE
RECEIVER

VCR

TV

Thomson Consumer Electronics

242

■ **10-20** *A plan for hookup of the receiver (DRD203RW) to a VCR and television.*

accurately pointed at the satellites. This makes it impossible to accurately point the DSS dish just by looking at the quality of the TV picture.

The DSS dish has two positioning adjustments. These adjustments are azimuth and elevation. The azimuth adjustment is the side-to-side movement of the dish. This is done by rotating the dish on the mounting post or mast. The elevation adjustment is done as an adjustment on the LNB support arm as shown in Fig. 10-21. This ad-

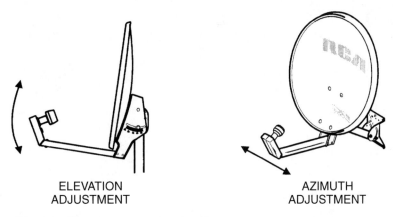

ELEVATION
ADJUSTMENT

AZIMUTH
ADJUSTMENT

■ **10-21** *Elevation and azimuth adjustments.* Thomson Consumer Electronics

justment has a reference scale on the side of the dish that is calibrated in degrees. If the mast is correctly plumbed, this scale will
be accurate. Detailed steps for aligning the satellite dish are given
in chapter 4.

Permission for use of information used in this chapter granted by
Thomson Consumer Electronics (RCA).

243

Appendix: Acronyms

AGC Automatic gain control; a circuit that provides for a fixed output, compensating for varying input levels.

AM Amplitude modulation; not used in satellite communications.

APS Antenna Positioning System, a mechanical/electronic system used in pointing the antenna at the satellite. For inclined-orbit satellites, it could also include a computer to anticipate the movement of the satellite.

BER Bit error rate. A digital circuit's performance is often measured in terms of a BER, or the probability of a digital bit being received correctly.

BO Backoff of the power amplifier. The value is commonly given as a negative number, representing an amount of power reduction required to avoid saturating the input to a satellite transponder.

BPSK Binary phase shift keying, a form of phase modulation in which two phases represent binary levels.

C/N Carrier-to-noise ratio, a value expressed in dB; good satellite links have a C/N of 10 dB or more.

CDMA Code division multiple access, a spread spectrum scheme transmitting on a bandwidth much larger than needed, but at a much lower level. PN (pseudorandom noise) code is used for retrieving the information.

CONUS Continental United States (contiguous 48 states)

CW Continuous wave, a signal at specific frequency that does not vary.

dB Decibel, used in various satellite expressions, provides for an easy way to write and understand both very small and very large ratios.

dBi Antenna gain in decibels, relative to an isotropic antenna; the true term to express the gain of a satellite antenna; i.e., a specific antenna has a gain of 43 dBi.

DBS Direct Broadcast Satellite, also known as DTH (Direct to Home) or TVRP.

dc Direct current, the kind of current you get from a battery.

DSBSC Double sideband suppressed carrier, an outmoded modulation scheme (see AM).

EHF Extremely high frequency, 30 to 300 GHz

EIRP Effective isotropic radiated power, stated as dBi when referring to the effective gain of an antenna.

EOL End of life, in regard to a satellite's useful lifetime.

f/D Focal distance to diameter ratio, the ratio of feedhorn distance to the center of the antenna divided by the diameter of the antenna.

FDM Frequency division multiplex, a simple approach to multiplexing in which each signal is assigned a specific frequency and bandwidth.

FDMA Frequency division multiple access. Simply put: Putting more than one signal through a transponder at a time, i.e., a 72-MHz transponder shared by two 36-MHz video carriers.

FEC Forward error correction, used with a digital signal to send the correction information before the actual signal data is sent.

FM Frequency modulation, a scheme in which the signal modulates the frequency of the carrier; the predominant transmission method for satellite communications.

FSK Frequency shift keying, in which binary levels are expressed with a shift in frequency.

FSS Fixed satellite service, a satellite service in which the ground station is at a fixed location (unlike MSS, where the earth station is mobile).

G/T The ratio of antenna gain (G) to the antenna system noise (T); a figure of merit, the higher the better; typical G/T specs are in the 20s and 30s dBi/K.

GEO Geostationary or geosynchronous orbit, in which the satellite is stationary in respect to the earth.

GHz Gigahertz (1,000 MHz)

GMT Greenwich Mean Time; the time in Greenwich, England, used as a standard time reference around the world.

Hemi Hemisheric beam, a term used when describing the footprint of satellite transponder.

HF High frequency, 3 to 30 MHz

HP Horizontal polarization, 90 degrees from vertical polarization; linear polarization.

I/O Input/output

IBS INTELSAT (see below) Business Services

IM Intermodulation noise or distortion, the undesirable result when a nonlinear amplifier is forced to carry multiple signals simultaneously.

INTELSAT International Telecommunications Satellite Organization

ISL Intersatellite link, an ITU definition for a radio communication link between satellites.

ITU International Telecommunications Union

LF Low frequency, 30 to 300 kHz

LHCP Left-hand circular polarization, equal components of horizontal and vertical linear signals.

LNA Low-noise amplifier

LNB Low-noise block down-converter

LNC Low-noise converter

LO Local oscillator

MCPC Multiple channel per carrier

MF Medium frequency, 300 kHz to 3 MHz

MSS Mobile Satellite Service

MTBF Mean Time Between Failures

MUX Multiplexer, a device that allows more than one signal within a specified bandwidth; the bandwidth could be divided by time, frequency, or a combination of both.

NF Noise figure, defined as the ratio between the input noise power to the output noise power, in comparison of an amplifier or amplifier stage.

NiCd Nickel cadmium battery

NiH Nickel hydrogen battery

PAL Phase Alternation by Line, the TV color system used primarily in most countries.

PCM Pulse Code Modulation

PFD Power flux density; a signal's illumination level as a measure of power received per unit area.

PI Polarization isolation; normally 30 dB between horizontal and vertical polarizations.

PM Phase modulation

PN Pseudorandom noise, as used in CDMA modulation schemes.

PSK Phase shift keying

PWM Pulse width modulation

QPSK Quadrature phase shift keying; a form of phase modulation similar to BPSK, except that there are four binary phase levels rather than two.

RF Radio frequency

RHCP Right-hand circular polarization

RMA Random (or contention) multiple access, a multiple access scheme in which no master station is used to control earth station transmissions. Upon transmission, the downlink signal is compared to a control signal. No corruptions leads the receiver to assume a good send; corruptions lead the receiver to assume conflict, and the packet is sent again.

RMS Root mean square

S/N Signal to noise ratio, a comparison of how much signal there is to how much noise.

SCPC Single channel per carrier; in digital transmission, this term is used to indicate how many signals are transmitted per digital carrier.

SDMA Space division multiple access

SECAM Systeme Electronique Coleur Avec Memoire, the TV color system of France, Russia, and some other countries.

SHF Super high frequency, 3 to 30 GHz

SSBSC Single sideband suppressed carrier

SSMA Spread spectrum multiple access

SSPA Solid-state power amplifier; an output amplifier used in satellite communications. Of the three types of amps, SSPAs provide the least power output and lower efficiency, are the most reliable, and have a very long life. Another type is the Klystron, which offers very high output with restrictions in heat dissipation and power sources.

TDMA Time division multiple access, in which the full transponder is used by one signal at a time, but on a shared-time basis. When one signal is on, the others wait. TDMA lends itself to digital signals very well because of the full-power capability and the use of burst sequences.

TI Terrestrial interference; interference caused by signals originating from the ground, usually a microwave signal sharing a common downlink frequency.

TVRO Television receive only, a common expression used to designate a backyard antenna system, BDS, DTH, etc.

TWT Traveling wave tube (see SSPA); An RF power amplifying device with very wide bandwidth (40 to 80 MHz) and good efficiency (30 to 50 percent); however, this device has a limited lifetime.

TWTA Traveling wave tube amplifier

UHF Ultrahigh frequency, 300 MHz to 3 GHz

VHF Very high frequency, 30 to 300 MHz

VLF Very low frequency, 3 to 30 kHz

VP Vertical polarization, linear polarization. This kind of polarization provides for doubling the useful bandwidth on adjacent frequencies, because it provides for 30 dB of isolation from horizontal polarization.

VSAT Very small aperture terminal; a small earth station with a small (about 1 to 2 meters) antenna.

VSWR Variable standing wave ratio

WARC World Administrative Radio Conference

Glossary

access card The access card is your DSS "credit card." It identifies you to the DSS service providers, and is required for your DSS box to function optimally.

address The unique identifier of a terminal on a network.

addressability The ability of a network, especially a satellite or cable system, to individually address and thus control (usually to enable decryption) the user's receiver.

adjacent channel An adjacent channel is one which is immediately next to another channel. For example: channels 2 and 3 and channels 8 and 9 are adjacent channels; channels 4 and 5 and channels 6 and 7 are not adjacent, as they have other frequencies allocated between them.

alignment bearing (rotor) A ball bearing-equipped guy ring that is slipped onto the antenna mast above the rotor to permit guying of the mast section rotated by the rotor.

alternate audio Alternate audio allows you to choose among the various audio options.

amplifier A device used to increase an electronic signal level.

amplitude modulation (AM) Also known as ancient modulation, not a modulation scheme used in satellite communications.

analog A signal that conveys information by means of changes in signal intensity and frequency. Fidelity is achieved by attaining a high enough signal-to-noise (S/N) ratio.

antenna A parabolic dish designed to collect electromagnetic signals from a satellite.

antenna discharge unit (lightning arrestor) A small device that is inserted into the transmission line and connected to a grounding wire or strap to discharge static electricity to ground before it can enter and damage a TV receiver.

antenna positioning system (APS) A mechanical/electronic system used in pointing the antenna at the satellite. For inclined orbit satellites, it could also include a computer to anticipate the movement of the satellite.

aperture A cross-sectional area of the antenna actually exposed to the satellite signal.

apogee The point in an elliptical satellite orbit that is farthest from the surface of the Earth.

attenuation The process of reducing the level of a signal either intentionally by means of an attenuator or by a natural process. A decrease in the strength (level) of a signal as it is transmitted or carried by wire(s) from one point to another. In antenna systems, attenuation is usually an undesirable characteristic.

attenuator A passive electronic device to reduce signal strength by a specified amount (usually expressed in dB).

attractions Attractions are special programs sent to you by your service provider.

automatic frequency control (AFC) A circuit that locks the receiving unit on to the frequency chosen and prevents drifting from that frequency.

automatic gain control (AGC) A circuit that boosts or retards gain to compensate for temporary gain shifts in the system.

az/el mount Antenna mount that requires two separate adjustments to move from one satellite to another; Azimuth and Elevation.

azimuth angle The angle of rotation (horizontal) that a ground-based parabolic antenna must be rotated through to point to a specific satellite in a geosynchronous orbit. The azimuth angle for any particular satellite can be determined for any point on the surface of the Earth, given the latitude and longitude of that point. It is defined with respect to due north, for convenience.

backmatch Backmatching refers to providing an accurate output impedance match (usually 75 Ωs in a TV system) from a piece of equipment such as a distribution amplifier or splitter. If the following load has a poor input impedance match, signals reflected from the load will be absorbed and not again reflected. Thus backmatching prevents multiple reflections, which cause ghosting or ringing.

backoff (BO) Backoff of the power amplifier to avoid saturating the input to a satellite transponder.

balun (matching transformer) A small device that matches the impedance of one component, transmission line, or circuit to that of another to prevent loss of signal strength and other unwanted characteristics. In antenna systems, baluns typically are used to match 75-Ω coaxial cable to the 300-Ω output of an antenna or the 300-Ω input of a TV.

band separator A device used to split two or more frequency bands into separate leads at the back of the television set. In MATV, they are normally UHF, VHF, and FM.

band, high (TV) See high-band TV.

band, low (TV) See low-band TV.

bandpass filter An active or passive device or circuit that passes a specific group of frequencies.

bandwidth A measure of spectrum (frequency) use or capacity. For instance, a voice transmission by telephone requires a bandwidth of about 3000 cycles per second (3 kHz). A TV channel occupies a bandwidth of 6 million cycles per second (6 MHz). Cable system bandwidth occupies 50 to 300 MHz on the electromagnetic spectrum.

baseband The basic direct 6-MHz output signal from a television camera, satellite television receiver, or videotape recorder. Baseband signals can be viewed only on studio monitors. To display the baseband signal on a conventional television set, a remodulator is required to convert the baseband signal to one of the VHF or UHF television channels that the TV set can be tuned to receive.

beamwidth The angle formed by the two compass directions that outline the boundaries of the area from which the front of an antenna can intercept signals and deliver them to the output at relatively equal levels. Generally, the narrower the beamwidth of an antenna, the greater the directivity and gain.

binary phase shift keying (BPSK) A form of phase modulation in which two phases represent binary levels.

bird Nickname for a satellite.

bit error rate (BER) In a digital system, corruption of the digital information is known as bit error. When the BER reaches a certain level, reception of the digital transmission will fail completely.

block down-converter A frequency band converter that changes higher frequency bands to lower frequency bands. A low-noise block down-converter changes the higher frequency satellite transmitting band (C-band or Ku-band) at its input to a lower intermediate-frequency band (IF band) at its output.

boot, weather See weather boot.

broadband Referring to a bandwidth greater than the baseband bandwidth, or greater than a voice frequency bandwidth. A device that will handle more than one channel.

C-band The 3.7 to 4.2 GHz band of frequencies, which is the dominant mode of satellite broadcast to the home dish owner.

carrier-to-noise ratio (C/N) C/N ratio is the difference in amplitude between the desired RF carrier and the noise present in that portion of the spectrum occupied by the desired signal. It is expressed in decibels to denote the ratio or relationship between the two. Usually measured in dB at the LNB output.

circular polarization A mode of transmission in which signals are downlinked in a rotating corkscrew pattern. A satellite's transmission-capacity can be doubled by using both right-hand and left-hand circular polarization.

Clarke belt The name given to the orbit 22,300 miles above the equator in honor of Arthur C. Clarke. Satellites placed in this orbit travel at the same apparent rate as the Earth's rotation, thereby maintaining a stationary position relative to a point on the Earth's surface. Generally, it is called the geostationary or geosynchronous orbit.

coaxial cable (coax) A type of round transmission line composed of a central conductor (wire) surrounded in turn by an insulating center (dielectric) and a metallic shielding material that typically is braided and acts as the second conductor. (Some types of coaxial cable have an aluminum foil shield under which is run a separate wire strand that serves as the second conductor.) These elements, in turn, are covered by a thick layer of insulating and weatherproofing material such as polyvinyl. Coaxial cables used as home TV transmission lines have an impedance of 75 Ωs.

code division multiple access (CDMA) A spread spectrum scheme, transmitting on a bandwidth much larger than needed, but at a much lower level. PN (pseudorandom noise) code is used for retrieving the information.

collocation The placing of several satellites near each other in orbit. Collocation allows a single fixed receiving antenna to receive signals from all the satellites without moving from one satellite to another (tracking).

communications satellite An artificial satellite, usually placed in geostationary orbit, used to relay radio transmissions.

continuous wave (CW) A signal at specific frequency that does not vary.

conus The contiguous United States; i.e., all the states (including the District of Columbia) except Alaska and Hawaii.

converter A device for changing signals from one frequency (or channel) to another frequency. (Ex., UHF Ch. 43 to VHF Ch. 7).

cross modulation A condition in which a strong signal will overload an amplifier and cause interference on other channels. Modulation that crosses over from one carrier to another. One type of "third order" distortion that occurs in all amplifiers as the signal levels are increased.

cross polarization The resulting condition when an LNB is located between horizontal and vertical polarization, or when isolation between left-hand circular and right-hand circular polarization is poor.

crosstalk The unwanted leakage of signal between supposedly independent channels.

dBi Antenna gain in decibels, relative to an isotropic antenna; the true term to express the gain of a satellite antenna, i.e.: a specific antenna has a gain of 43 dBi.

dBM The ratio in dB of the power of a signal as compared with a one-milliwatt reference power.

dBMV The ratio in dB of the power of a signal compared to one millivolt in a 75-Ω system.

dBW The ratio of the power to one watt, expressed in decibels.

decibel (dB) A measurement term that describes the strength (level) of a signal in logarithmic relation to a reference (level). For example, when the signal is expressed in micro-volts, an increase of the signal by 6 dB means that the signal strength has been doubled.

declination angle Sometimes called the "offset angle." The TVRO dish uses a polar mount to track all satellites on the geostationary orbit. In order to point the dish at a satellite 22,300 miles above the equator, the dish is first included skyward so its pivot points are perpendicular to the plane of the equator. This makes the dish look parallel to the equator's plane. The reflector, and not the pivot points, is then declined slightly to look down to the orbiting satellites. As the polar mount tracks across the sky, it will stay aligned with the geosynchronous orbit.

decoder A device that reconstructs an encrypted signal so that it can be clearly received.

demodulation The process of extracting the original signal from a modulated carrier.

demodulator A device that separates the information signals from the carrier.

255

detector A demodulation circuit used in a satellite television receiver to recover the audio and video signals from the carrier.

dielectric An insulating material placed between conductors to prevent the conductors from physically contacting one another (shorting out). In coaxial cable, insulating material surrounds the center conductor to prevent it from touching the metallic shield (and other conductor wire if one is used). The insulating material also maintains a specific amount of space between the center conductor and the other conductor. This spacing is necessary to maintain certain cable characteristics that if changed, will decrease the quality of the signal.

digital Conversion of information into bits of data for transmission through wire or over air. Method allows simultaneous transmission of voice and data.

digital audio Method used to transmit audio on scrambled channels.

digital signal A signal that can take on only certain, discrete values.

dipole The element(s) of an antenna that intercepts the signal and feeds it to the antenna output terminals. The other elements of the antenna serve as "director" and "reflector" that direct or reflect the incoming signal to the dipole element.

direct broadcast satellite (DBS) Refers to service that uses satellites to broadcast multiple channels of television programming directly to small-dish antennas.

direct current (dc) The stuff you get from a battery.

directivity The ability of an antenna to pick up signals from one general direction (usually from the front) and effectively reject those from other directions (usually from the back and sides). The front-to-back ratio is one measure of an antenna's directivity.

DirecTV DBS subsidiary of Hughes Communications. Responsible for system management and programming. DirecTV and Hughes are principal driver and partner in 101 W.L. high powered DBS project.

discharge unit, antenna See Antenna Discharge Unit.

dish Nickname for a parabolic satellite antenna.

dish illumination That portion of the dish the feed actually sees.

distribution amplifier An amplifier that is mounted indoors to boost the strength (level) of the received signal so that it can be fed to two or more receivers.

distribution system That part of an MATV system that carries the signals from the head end to the individual TV sets.

dither Appears as a 30 Hz flickering in the picture, usually caused by insufficient removal of the 30 Hz dispersion waveform transmitted with the video from the satellite. Intention is to prevent the satellite signal from interfering with terrestrial microwave links.

double sideband suppressed carrier (DSBSC) An outmoded modulation scheme (see AM).

downlink Term used to describe the retransmitting of signals from a satellite back to Earth.

drip loop A short, U-shaped loop of a wire (or cable) immediately adjacent to a house entry point or electrical connection, so that water will drain off of the wire and not run into the house or connection.

DSS Trademark of Hughes Communications for hardware used to receive high powered DBS from 101 west longitude satellites.

earth station The electronic ground equipment used in conjunction with a parabolic shaped antenna for receiving and processing radio frequency signal and/or transmitting them to a satellite.

effective isotropically radiated power (EIRP) A measure of the effective power emitted by a transmitter, or a measure of the signal strength received on Earth from a satellite. Also a measure of the power of a satellite television signal received on Earth and expressed in dBW.

element, antenna The small, hollow metal rods of various lengths that are attached (usually perpendicularly) to the main horizontal support member (boom or crossarm) of the antenna. The element at the rear of the antenna (called reflector) is usually the longest. The element that actually feeds the intercepted signal to the antenna output is called a dipole.

elevation Elevation refers to the up and down positioning of your DSS dish. When you enter your zip code (or latitude and longitude), you will be given a number corresponding to an elevation number required for your geographic location.

encoder (scrambler) A device used to electronically alter a signal so that it can only be viewed on a receiver equipped with a special decoder.

encryption Altering a program for the purpose of controlling access. In television, a program is encrypted with a scrambler at the source of transmission. A descrambler is required to decrypt the program.

end of life (EOL) In regard to a satellite's useful lifetime.

equalization The technique of compensating for differences in attenuation of a signal at different frequencies.

equivalent satellite link noise temperature The noise temperature of the receiving antenna of the earth station corresponding to the radio frequency noise power that produces the total observed noise at the output of the satellite link excluding noise due to interference coming from satellite links using other satellites and from terrestrial systems.

extremely high frequency (EHF) 30 to 300 GHz

F-connector A small, metallic, male-type connecting device with internal threads that attach to the end of a coaxial cable to secure and electrically connect the coax to a female F-fitting. The internal threads of the male connector screw onto the external threads of the female connector. Most baluns have a female-type F-connector on one end for the 75 Ω coax, and terminal lugs on the other end for 300 Ω twinleads.

feedhorn A device that gathers microwave signals reflected from the surface of the dish and feeds them to the LNB.

feedline The transmission line, typically coaxial cable, from the satellite antenna to the receiver.

field strength meter An electronic instrument that measures the strength of a signal and indicates it on a meter usually calibrated in dBmV.

filter Filters are used to block out undesired frequencies. There are two types of filters: bandpass and rejection. A bandpass filter permits only the desired range to pass through, while a band reject filter attenuates an undesired range of frequencies.

fixed satellite service (FSS) a satellite service in which the ground station is at a fixed location, unlike MSS where the earth station is mobile.

FM threshold That point at which the input signal power is just strong enough to enable the receiver demodulator circuitry successfully to detect and recover a good quality television picture from the incoming video carrier.

focal-distance-to-diameter ratio (F/D) Focal distance to diameter ratio, the ratio of feedhorn distance to the center of the antenna divided by the diameter of the antenna.

focal length Distance from the feed to the center of the antenna reflector.

focal point The area toward which the primary reflector directs and concentrates the signals received.

footprint The geographic area toward which a satellite directs its signal.

forward error correction (FEC) Forward error correction, used with a digital signal to send the correction information before the actual signal data is sent.

frequency The property of an alternating-current signal measured in cycles per second or hertz. In general, the higher the frequency of a signal, the more capacity it has to carry information, the smaller an antenna is required, and the more susceptible the signal is to absorption by the atmosphere and physical structures. At microwave frequencies, radio signals take on a line-of-sight characteristic, and require highly directional and focused antennas to be seen successfully.

frequency-agile The ability of a satellite TV receiver to select or tune all channels (transponders) from a satellite. Receivers not frequency-agile are dedicated to a single channel, and are most often used in the CATV industry. Frequency agility can be via continuously variable tuning or discrete-step (channel selection) tuning.

frequency division multiple access (FDMA) Simply put— putting more than one signal through a transponder at a time, i.e.: a 72 MHz transponder shared by two 36 MHz video carriers.

frequency division multiplex (FDM) A simple approach to multiplexing in which each signal is assigned a specific frequency and bandwidth.

frequency modulation (FM) A scheme in which the signal modulates the frequency of the carrier; predominant transmission method for satellite communications.

frequency shift keying (FSK) In which binary levels are expressed with a shift in frequency.

front-to-back ratio A measure of the directivity of an antenna that is based on the difference between the strengths of signals received from the antenna front and those received from the back. The difference usually is expressed in decibels (dB). For example, a front-to-back ratio of 40 dB indicates that the power of signals received from the antenna front will be 10,000 times greater than those received from the back. Generally, the higher the rating in dB, the greater the directivity of the antenna.

gain (satellite antennas) The ratio of radiated power of a given antenna to that of a lossless isotropic antenna (radiates equally in all directions.) Usually expressed in decibels relative to isotropic (dBi) and in the direction of maximum power.

gain (UHF, VHF, FM antennas) The ratio of radiated power of a given antenna to that of a lossless half-wave dipole antenna. Usually expressed in decibels (dB) relative to a dipole (dBd) and in the direction of maximum power.

gain compression The gain of any amplifier is a function of the input signal level, and will always decrease at some point where the input level is sufficiently high. The capacity of an amplifier is sometimes expressed as the level where the gain of the amplifier is decreased or compressed by a certain amount. This is especially true of a single frequency amplifier where cross-modulation is not meaningful.

gain-over-noise temperature (G/T) The ratio of antenna gain (G) to the antenna system noise (T); a figure of merit, the higher the better; typical G/T specs are in the 20s and 30s dBi/K.

geostationary See Geosynchronous.

geosynchronous orbit (GSO) An orbit around the Earth with an average distance from the center of Earth of about 26,000 miles, in which a satellite would have a period equal to the rotation period of the Earth. Most communications satellites are in geosynchronous orbit.

ghosts (ghosting) Faint duplicate images that appear in a TV picture to either the left or right of the desired picture image.

gigahertz (GHz) A frequency equal to one billion Hertz, or cycles per second. See frequency or Hertz.

greenwich mean time (GMT) Greenwich Mean Time (like in England)

ground rod A long metal rod that is driven into the ground near an antenna installation and to which is attached the grounding wires from the mast and antenna discharge unit to discharge static electricity to ground before it can enter and damage the TV receiver.

guard channel An unused channel space that serves to prevent the nearby television channels from interfering with each other.

Guide Guide lists the programs available from your service provider.

guy ring A circular metal collar with attachment holes (eyes), that is slipped on and clamped to an antenna mast. Guy wires are then attached to the mast through the holes in the guy ring.

guy wire (guying) Three or more multi-strand steel or aluminum wires that are connected between the guy ring(s) on the antenna mast and widely spaced eye screws in the house roof, supporting the mast against the forces of wind and ice.

head end The master distribution center of a CATV system in which the incoming television signals from space and distant broadcast stations are received, amplified, and remodulated onto television channels for transmission down the CATV coaxial cable.

Hertz (Hz) The name of basic measure of frequency with which an electromagnetic wave completes a full cycle from its positive to its negative pole and back again. Each unit is equal to one cycle per second. See frequency.

high-band TV The band of frequencies assigned to VHF TV channels 7 through 13 by the Federal Communications Commission (FCC).

high-definition television (HDTV) A television picture of higher quality than standard NTSC video. This usually involves more horizontal lines. (e.g., 1050 rather than 525 and a different aspect ratio-16:9 rather than 4:3)

high-frequency (HF) 3 to 30 MHz

high-pass filter A device that is connected to a transmission line to filter out interfering signals whose frequencies are below those in the TV band. High-pass filters typically are used to filter out interference caused by amateur and citizens band radio transmission.

horizontal polarization (HP) Horizontal polarization, 90 degrees from vertical polarization, linear polarization.

HS-601 Hughes Satellite #601. The type of satellite launched in December 1993 and June 1994 to transmit DirecTV and USSB programming received on DSS hardware.

impedance A signal-affecting characteristic that is present to some degree in all electrical conductors (wires) and electronic circuits. Impedance is usually expressed in Ωs. To prevent an unnecessary decrease in the strength of a signal that is being transferred (coupled) from one type of conductor or circuit to another, the difference in impedance must be "matched" by a device that compensates for the differences in the impedances. A balun is used in antenna systems to compensate for the differences in impedance between a 300

261

Ω antenna and a 75 Ω coax, and between a 75 Ω coax, and the 300 Ω input circuit of a TV receiver. Impedances that are not the same, or whose differences have not been compensated for, are said to be mismatched.

inclination The angle between the Earth's equatorial plane and the orbit plane of the satellite (deg.).

input capability, preamp The maximum strength of signal that an antenna preamp can accept without "overloading." (Overloading causes distortion, reduction, or complete elimination of the signal.)

insertion loss Signal lost when a piece of equipment is inserted in the line, commonly referred to as "feed thru" loss.

integrated receiver-decoder (IRD) A term often used in referring to a Satellite Receiver. New satellite receivers feature a built-in decoder (VideoCipher II Plus or VideoCipher RS) and dish motor drive controller. In older models, these were separate components.

interaction, signal The interfering effect that one signal has on another when two different signals are present at the same time in a conductor (wire) or circuit. Signal interaction in the transmission line of a TV antenna system causes picture problems such as ghosting, smearing, snow, and various forms of interference patterns.

interference A jumbling of the content of signals by the reception of undesired signals.

intermodulation (IM) Noise or distortion, the undesirable result when a nonlinear amplifier is forced to carry multiple signals simultaneously.

intersatellite link (ISL) An ITU definition for a radio communication link between satellites.

isolation The amount of separation or loss (expressed in dB) between two locations or components. (Such as the loss between the feed-thru and tap/drop line of a tapoff unit.)

K- (or Ku-) band Used by a new generation of high powered satellites and smaller dishes, including high powered DBS. The 12.2 to 12.7 GHz frequency band reserved for the exclusive use of broadcast satellites.

kelvin (K) The temperature measurement scale used in the scientific community. Zero degrees Kelvin represents absolute zero, and corresponds to minus 459 degrees Fahrenheit. Thermal noise characteristics of an LNB are measured in degrees Kelvin, with respect to room temperature. (See Noise Temperature.)

key Key is a 4-digit password that allows you to lock or unlock the various features of your system.

kiloHertz (kHz) A unit of frequency equal to 1,000 Hertz. See Frequency or Hertz.

latitude An angular measurement of a point on the Earth above or below the equator. The equator represents zero degrees, the north pole plus 90 degrees, and the south pole minus 90 degrees.

left-hand circular polarization (LHCP) Equal components of horizontal and vertical linear signals.

limits There are 2 kinds of limits. The Ratings Limit allow you to control program content by ratings level. The Spending Limit controls spending on a program by program basis.

line amplifier An amplifier inserted in a transmission line to compensate for the losses in that line.

line splitter A device that divides the signal from a single cable into two or more cables of similar impedance.

locks The lock prevents access to various features of your system. The lock is controlled by a 4-digit key that acts as a password.

longitude An angular measurement of a point on the surface of the Earth in relation to the meridian of Greenwich (London). The Earth is divided into 360 degrees of longitude, beginning at the Greenwich mean. As one travels west around the globe, the longitude increases.

look-angle The angle of the dish above the local horizontal when the dish is looking at a satellite.

low-band TV The band of frequencies assigned to VHF TV channels 2 through 6 by the Federal Communications Commission (FCC).

low frequency (LF) 30 to 300 kHz

low noise amplifier (LNA) A preamplifier used at the feedhorn of the satellite antenna to amplify the weak satellite signal.

low noise block A device that amplifies and down-converts the down-converter (LNB) whole 500 MHz satellite bandwidth at once to an intermediate frequency range.

low noise block down-converter (LNB) A microwave assembly attached to the feed's polarizer. It includes an LNA and a block down-converter. (See Block Down-converter).

magnetic deviation The angular adjustment, expressed in degrees, between a magnetic compass indication and the true, or geodetic indication.

mailbox The Mailbox holds your incoming messages and reminders. The Mailbox is accessed through the on-screen menu system.

main menu The main menu lists an overview of the features available through the on-screen menu system.

mast, antenna (TV) A vertical section (or sections) of tubular steel or aluminum on which the antenna is mounted. Most sections typically are available in 5 and 10 ft. lengths.

master antenna television (MATV) The master distribution coaxial cable television antenna system found in a modern apartment building, condominium complex, or hotel.

matching impedance See Impedance.

matching transformer (TV) See Balun.

medium frequency (MF) 300 kHz to 3 MHz

megaHertz (MHz) Refers to a frequency equal to one million Hertz, or cycles per second. See Frequency or Hertz.

microvolt (μV) One millionth of a volt, or 0.000001 volt. The strength of the signals in a TV antenna system is expressed as so many microvolts (μV). Generally, to produce an acceptable TV picture, the strength of the TV signals at the input terminals of a TV set must be at least 1000 microvolts (μV).

microwave Line-of-sight, point-to-point transmission of signals at high frequency. The portion of the electromagnetic spectrum above 1 GHz in frequency.

mid-band The part of the frequency band that lies between television channels 6 and 7, reserved by the FCC for air, maritime and land mobile units, FM radio and aeronautical and maritime navigation. Mid-band frequencies, 108 to 174 MHz, and also be used to provide additional channels on cable television systems.

mixer A device for changing the frequency of a signal.

modulation The process of adding information such as video and/or audio to an RF carrier.

modulator A device that modulates a carrier. Modulators are found as components in broadcasting transmitters and in satellite transponders.

Motion Picture Experts Group (MPEG) This international group will create a worldwide "standard" for video compression.

multiple array Two or more antennas mounted on the same mast with outputs coupled together. Multiple arrays are used to increase gain and directivity.

264

multiplexer (MUX) Provides for a method to accommodate more than one signal within a specified bandwidth, which could be shared by time, frequency or a combination of both.

National Rural Telecommunication Cooperative (NRTC) Association represents 2400+ rural power and telephone co-ops. Dedicated to bring cable type TV programming to rural Americans via satellite. Investing $250 million in DirecTV to be a partner.

noise Any unwanted contribution to a signal. May be natural or interference from other signals.

noise figure Generally, a numerical rating that indicates how much electromagnetic "noise" there will be at the output of a circuit or system compared to the noise at the input. If the strength of the noise is too near that of the picture-producing signal, noise-produced specks, called "snow," will be produced on the TV screen. The lower the noise figure of a circuit, system, or component, the lower the output noise level will be compared to the output signal level and in turn, the lower the probability of snow on the screen of the TV set.

noise temperature The noise temperature of a low-noise amplifier (LNB) is expressed in degrees Kelvin with respect to room temperature.

off-the-air The reception of a TV signal that has been broadcast over the air without the assistance of microwave relay.

offset feed A specialized feed offset from the actual antenna focus point but is at the theoretical focus point of the parabolic antenna from which the actual antenna is taken.

Ω The unit of measure of resistance and impedance (See Impedance).

omnidirectional (antenna) An antenna capable of receiving signals from all compass directions equally well. Such an antenna is nondirectional.

orbit The path, relative to a specified frame of reference, described by the center of mass of a satellite or other object in space subjected primarily to natural forces, mainly the force of gravity.

orbit altitude Height above the Earth's surface (km).

orbit period The time for the satellite to make one full revolution in its orbit (min).

orbit spacing The angular separation (measured in degrees of longitude) between satellites using the same frequency and covering overlapping area (deg.).

orient (an antenna) To aim the antenna in a specific direction, usually toward the transmitting tower(s) of the TV stations.

overloading of preamp See Input Capability, Preamp.

overloading of receiver See Receiver Overload (TV).

parabolic antenna The most frequently found satellite TV antenna. It takes its name from the shape of the dish described mathematically as a parabola.

passive device Any signal-handling device that is not electrically powered and therefore, does not increase the strength of the signal. Couplers and splitters are examples of passive devices.

pay-per-view (PPV) A system whereby a cable or DBS customer can order individual programs, such as films or sporting events, either immediate or for some specific later time.

perigee The point in an elliptical satellite orbit that is closest to the surface of the Earth.

period (of a satellite) The time elapsing between two consecutive passages of a satellite through a characteristic point on its orbit.

phase alteration by line (PAL) The TV color system used primarily in most countries.

phone prefix Certain phone systems require a single-digit phone prefix (such as "9") that must be dialed before you can reach an outside line.

picture carrier (TV) The part of a TV signal that contains the video (picture) information. The audio (sound) is contained in the sound carrier.

pitch pad A small piece of neoprene or other "rubbery" material that is placed under the legs of a tripod roof mount to cushion the mount and seal around the anchor bolts that secure the legs to the roof.

plumb line A plumb line is a weight attached to a string. The weighted string provides a reference line perpendicular to the ground.

point and select Point and select is the method of using the remote control to highlight different parts of the on-screen display.

polar mount Antenna mount used to allow only one required movement to point the antenna to any satellite in the Clarke Belt. (As opposed to AZ/EL.)

polar plot A flat graph that provides a bird's-eye-view of antenna performance characteristics such as directivity and beamwidth.

polarization The directional aspects of a signal. Signals can have circular or linear polarization.

polarization isolation (PI) Normally 30 dB between horizontal and vertical polarizations.

power The strength of a signal as measured in watts.

power flux density (PFD) Power flux density, illumination level as a measure of power received per unit area.

preamp, TV (preamplifier) A small amplifying device that is mounted on the mast or antenna boom as close to the antenna output terminals as possible, so that the strength of a very weak signal is increased (amplified) before it enters the transmission line. Without this preamplification, the strength of the already weak signal (1000 μV or less) would be further reduced as it passes through the transmission line producing "snow" on the TV screen.

prime focus Term given to the types of antennas that focus signals to a primary focus point.

program guide The program guide lists the programs available from your service provider.

pseudorandom noise (PN) Pseudorandom noise, as used in CDMA modulation schemes.

purchase history Purchase history lists information on programs that you have already viewed.

quadrature phase shift keying (QPSK) A form of phase modulation similar to BPSK, but there are four binary phase levels.

radio frequency interference (RFI) Interference with a receiver by a signal inadvertently generated at or near the frequency that the receiver is attempting to receive. The Federal Communications Commission establishes RFI standards for electronic devices (such as video games) to minimize this interference.

radio-frequency spectrum Those frequency bands in the electromagnetic spectrum that range from several hundred thousand cycles per second to several billion cycles per second.

rain fade (loss) The attenuation of a signal due to rainfall. It should be noted that the noise temperature perceived by the receiving antenna may also increase due to rain being present in the link.

random multiple access (RMA) Random (or contention) multiple access, multiple access scheme in which no master station is used to control earth station transmissions. Upon transmission, downlink signal is compared, no corruptions assume a good send; corruptions assume conflict and packet is sent again, at random.

rating limit The rating limit allows you to control program content by ratings level.

rear rejection The ability of an antenna to reject (not receive) signals that approach it from the back.

receiver An electronic device that detects and decodes a radio or television signal.

receiver isolation The attenuation between any two receivers connected to the system.

receiver overload (TV) A condition in which excessively strong signals cause the picture on the TV receiver to become distorted. Older model receivers are more likely to overload than are newer ones. Frequently, this condition can be eliminated by readjusting the receiver's automatic gain control (AGC).

receiver sensitivity A measure of how much input signal is necessary to achieve a specified base band performance, such as a specified bit-error rate or signal to noise ratio.

repeater Any device that receives a signal and retransmits it to parts of a network farther down the line, usually amplified or restored to proper shape and intensity. On a communications satellite, a transponder.

return loss The ratio of power incident on a load to power reflected. The reflected signal occurs when the source and load are mismatched.

RG-6 75-Ω coaxial cable.

satellite A sophisticated electronic communications relay station orbiting 22,300 miles above the equator moving in a fixed orbit at the same speed and direction of the Earth (about 7,000 mph east to west).

satellite master antenna television (SMATV) A cable television system, sometimes serving a large area, but more commonly referring to service to an apartment building, housing complex, etc.

satellite receiver A digital or wideband FM receiver operating in the microwave range, converting the incoming C-band or Ku-band RF signal to a standard baseband video signal.

satellite terminal A receive-only satellite earth station consisting of an antenna (typically parabolic in shape), a feedhorn, a low-noise amplifier, a down-converter, and a satellite receiver.

scrambling Any system intended to render a received message unintelligible or unviewable without authorization. Also called encryption and encoding.

Sensitivity, antenna General classifications of relative antenna gain that indicate approximately how far from the station transmitter tower(s) an antenna is designed to be used. Examples of these classifications are suburban, fringe, and deep fringe.

side lobes Energy pickup offset from the main focus direction of the antenna.

signal meter The signal meter displays the relative strength of the satellite signal. The signal meter is especially useful when you adjust your DSS dish for the first time.

signal mismatch A condition in which an antenna system delivers signals whose strengths and general quality vary. This usually is the result of incorrectly installed signal-distribution components. (See Interaction, Signal.)

signal-to-noise ratio (S/N) Signal to noise ratio, a comparison of how much signal there is to how much noise.

single channel per carrier (SCPC) Single channel per carrier, in digital transmission, used to indicate how many signals are transmitted per digital carrier.

slot The longitudinal angular position in the geosynchronous orbit into which a communications satellite is parked.

smear A term used to describe a picture condition that objects appear to be extended horizontally beyond their normal boundaries in a blurred or "smeared" manner.

snow A form of noise picked up by a television receiver, caused by a weak signal. Snow is characterized by alternate dark and white dots randomly appearing on the picture tube.

solar outage If an antenna is pointed at or near the sun, the high radiated noise level of the sun may be many times stronger than the desired signal. Solar outages occur when an antenna is looking at a satellite, and the sun passes behind or near the satellite and within the field of view of the antenna. This field of view is usually wider than the beam width. Solar outages occur twice a year and are predictable as to the timing for each site.

solid-state power amplifier (SSPA) Solid-state power amplifier, output amplifier used in satellite communications. Of the three types of amps, SSPAs provide the least power output and lower efficiency, is the most reliable and has a very long life. The third type is the Klystron, which offers very high output with restrictions in heat dissipation and power sources.

sound carrier (TV) The part of a TV signal that contains the audio (sound) information. (The picture information is contained in the picture carrier.)

spacing Length of cable between amplifiers expressed as dB loss at the highest TV channel in a system, with equal amplifier gain in main trunks.

sparklies A form of satellite television "snow" caused by a weak signal.

spending limit The spending limit controls spending on a program-by-program basis.

spherical antenna An antenna with the ability to simultaneously see and receive several satellites.

splitter A device with one input providing two or more outputs in the same frequency.

spot beam A focused, high power satellite signal that covers only a small region. Outside that area, the signal is undetectable, and will not interfere with other use of the same wavelength.

standout (standoff) A metallic device with woodscrew threads or a clamp on one end and a circular loop (eye) with slotted insulating material on the other. It is used to secure and hold 300 Ω twinlead or other unshielded transmission line away from metal gutters, walls and other surfaces that can change the line's signal-handling characteristics. The standout is screwed into a wall or other part of the house or is clamped onto the antenna mast. The transmission line then is inserted in the slot of the insulating material in the eye.

sub band The frequency band from 6 MHz to 54 MHz, which may be used for two-way data transmission.

subcarrier A second signal piggybacked onto a main signal to carry additional information. A carrier of a lower frequency modulating a main carrier with a higher frequency.

superband The frequency band from 216 to 600 MHz, used for fixed and mobile radios and additional television channels on a cable system.

super high frequency (SHF) Super high frequency, 3 to 30 GHz

switchable trap A small device that is used with a preamp to eliminate (trap out) an unwanted bank of signals. It is called "switchable" because it can be switched on or off (although this is difficult because the trap is mounted with the preamp up near the antenna). A switchable trap usually is used to eliminate the FM band.

synchronization The process of orienting the transmitter and receiver circuits so that information sent in relation to a precise instant of time by the transmitter will be perfectly related to that same instant by the receiver.

system test The system test is a feature that can be used to ensure that your DSS equipment is working properly.

Systeme Electronique Coleur Avec Memoire (SECAM) The TV color system of France, Russia and some other countries.

tandem system A preamp and a distribution amplifier that are designed to be used together.

tapoff A device that, when installed in the feeder line, will allow a specific amount of energy to be removed from the thru line and to be fed to a TV set.

television receive-only (TVRO) Trade name often used for satellite receiving system, shorthand for television receive-only. Home dish owners receive signals while programmers send or uplink signals.

tensile strength The ability of a material or structure (such as an antenna mast) to withstand large bending forces without distorting or breaking apart. Tensile strength is an important factor to consider when selecting a mast and planning an installation.

terminator A resistive device that is used to match the cable to its characteristic impedance. When installed at the ends of the thru line, it prevents reflections back down the line.

terrestrial interference Interruptions in a satellite signal caused by high power land-based microwave links in the 4 GHz band.

terrestrial interference (TI) Terrestrial Interference, caused by signals originating from the ground, usually a microwave signal sharing a common downlink frequency.

terrestrial tv Ordinary VHF (very high-frequency) and UHF (ultra high-frequency) television transmission limited to an effective range of 100 miles or less.

themes Themes are a type of "filter" that are based on entertainment categories. Themes can be used to make it easier to find information in program guide.

threshold extension A technique used by satellite television receivers to improve the carrier-to-noise threshold of the receiver by approximately 3 dB.

tilt control A device on a distribution amplifier that adjusts the amplification to compensate for the attenuation slope of the cable loss.

time division multiple access (TDMA) Time division multiple access, in which the full transponder is used by one signal at a time, but shared on a time basis. When one signal is on, the others wait. TDMA lends itself to digital signals very well because of the full power capability and burst sequences.

transfer orbit An intermediate orbit used to reach the geosynchronous orbit.

transmission line A two (or more) conductor wire that is used to carry current or signals from one point to another. Twinlead and coaxial cable are the most common types of transmission line used to carry TV signals from the antenna(s) to the receiver.

transmitter An electronic device consisting of an oscillator, a modulator and other circuits that produce a radio or television electromagnetic wave signal for radiation into the atmosphere by an antenna.

transponder A combination receiver, transmitter, and antenna package on a satellite. The signal is received from the uplink station and downlinked back to earth on a different frequency. Most C-band satellites have 24 transponders (12 vertical, 12 horizontal polarization); most Ku-bands satellites have 16, although the Canadian Anik birds have 32. Some satellites have a mix of C-band and Ku-band transponders.

trap A frequency sensitive device that is used to attenuate specific signals that cause interference.

traveling wave tube (TWT) Very wide bandwidth (40 to 80 MHz), good efficiency (30 to 50 percent); has limited lifetime.

tuneable trap A small device that can be tuned (adjusted) to eliminate any one of the number of frequencies within a band. Tunable traps are frequently used with preamps to eliminate a particularly troublesome signal.

tuner That portion of a receiver that can variably select, under user control, a desired signal from a group of signals in a frequency band.

TV/DSS TV/DSS is a button that switches (toggles) the source of the video signal from the DSS dish to a home antenna/cable. This button is similar to the TV/VCR button that some remote controls have.

TV/VCR The TV/VCR button switches the incoming signal source from the TV antenna to the VCR.

twinlead A type of unshielded ribbon-like transmission line that consists of two insulated conductors (wire) separated by a thin, flat expanse of insulating material. TV twinlead has a characteristic impedance of 300 Ωs, and therefore is called 300 Ω twinlead.

ultra high frequency (UHF) When used in relation to TV, UHF refers to channels 14 through 69, whose frequencies are located in the UHF band. Modern TV receivers have two separate tuners, one for VHF (channels 2 through 13), and one for UHF (channels 14 through 69). Although the UHF tuners of some television receivers can also be tuned to channels 70 through 83, there are no TV signals on these channels because the FCC has reassigned their frequencies to other users.

United States Satellite Broadcasting (USSB) A division of Hubbard Broadcasting, USSB owns part of the 1st Hughes Satellite and will sell 25–30 channels of programming available on DSS.

uplink A signal from an earth transmit station to a satellite, or an antenna facility that transmits signals to a satellite.

vertical interval test signal A method whereby broadcasters add test signals to the blanked portion of the vertical interval. Normally placed on lines 17 through 21 in both field one and two.

vertical polarization (VP) Vertical polarization, linear polarization, provides for doubling the useful bandwidth on adjacent frequencies; provides for 30 dB isolation from horizontal polarization.

very high frequency (VHF) When used in relation to TV, VHF refers to channels 2 through 13, whose frequencies fall within the VHF band. The TV VHF band is divided into two sub-bands: the low band, which includes channels 2 through 6 and a frequency range of 54 MHz–88 MHz, and the high band, which includes channels 7 through 13 and a frequency range of 174 MHz–216 MHz. A portion of the frequency band between channels 6 and 7 is used for FM radio stations.

very low frequency (VLF) Very low frequency, 3 to 30 kHz

very small aperture terminal (VSAT) A satellite dish less than 1.8 meters in diameter, typically located at a retail or other remote location as an endpoint of a VSAT network.

video The portion of television that is visual.

video compression A digital technique to compress several video channels into the bandwidth normally required by one channel.

voltage standing wave ratio (VSWR) A measurement of the difference between the minimum and the maximum voltage along the transmission line. The ideal VSWR is 1.0. As the VSWR increases so does the possibility of ghosting.

waveguide A metal tube used to guide or transmit microwave signals between two given points.

weather boot A rubber-like covering that is used to protect outdoor electrical connections from the weather (rain, ice, etc.).

wideband See Broadband

X-band Used loosely to refer to satellites operating in the 8/7 GHz range.

zenith The highest look-angle for a dish constrained only by its construction. A polar-mounted TVRO's zenith is a point on the Clark Belt due south of the dish, or a satellite located on the same longitude as the antenna.

Information courtesy of Thomson Electronics

Index

Illustrations are in **boldface**.

277

279

280

281

About the author

A respected author of technical electronics books and articles, Robert L. Goodman has been employed in the electronics service industry for more than 45 years. He is an electronics repair consultant and test instrument designer; holds amateur radio and General Radiotelephone Operator's licenses; and is a Certified Electronics Technician and Certified Engineering Technician. His more than 45 electronics books include *Maintaining and Repairing VCRs, Fourth Edition* and *Troubleshooting & Repairing Digital Video Systems*.

283